Laboratory Manual *for*

Soil Science

Agricultural & Environmental Principles

Eighth Edition

Stephen J. Thien
Kansas State University

John G. Graveel
Purdue University

Boston   Burr Ridge, IL   Dubuque, IA   Madison, WI   New York   San Francisco   St. Louis
Bangkok   Bogotá   Caracas   Kuala Lumpur   Lisbon   London   Madrid   Mexico City
Milan   Montreal   New Delhi   Santiago   Seoul   Singapore   Sydney   Taipei   Toronto

McGraw-Hill Higher Education
*A Division of The **McGraw-Hill** Companies*

LABORATORY MANUAL FOR SOIL SCIENCE: AGRICULTURAL & ENVIRONMENTAL PRINCIPLES, EIGHTH EDITION

Published by McGraw-Hill, a business unit of The McGraw-Hill Companies, Inc., 1221 Avenue of the Americas, New York, NY 10020. Copyright © 2003, 1997 by The McGraw-Hill Companies, Inc. All rights reserved. No part of this publication may be reproduced or distributed in any form or by any means, or stored in a database or retrieval system, without the prior written consent of The McGraw-Hill Companies, Inc., including, but not limited to, in any network or other electronic storage or transmission, or broadcast for distance learning.

Some ancillaries, including electronic and print components, may not be available to customers outside the United States.

 This book is printed on recycled, acid-free paper containing 10% postconsumer waste.

1 2 3 4 5 6 7 8 9 0 QPD/QPD 0 9 8 7 6 5 4 3 2

ISBN 0–07–242881–3

Publisher: *Margaret J. Kemp*
Sponsoring editor: *Thomas C. Lyon*
Freelance developmental editor: *Kassi Radomski*
Executive marketing manager: *Lisa Gottschalk*
Senior project manager: *Jill R. Peter*
Production supervisor: *Sherry L. Kane*
Coordinator of freelance design: *Michelle D. Whitaker*
Cover designer: *Rokusek Design*
Cover images: Hand/Plant image: *Dan Donnert, Kansas State University Photographic Services*
Soil images: *Soil Survey, Natural Resources Conservation Service, USDA*
Blue Marble image: *NASA Goddard Space Flight Center*
Compositor: *GAC Indianapolis*
Typeface: *12/14 Times Roman*
Printer: *Quebecor World Dubuque, IA*

www.mhhe.com

Contents

Preface

Exercises

1. Soil as a Natural Resource *1*
2. Soil Texture *23*
3. Particle Size Distribution *35*
4. Bulk Density and Soil Porosity *47*
5. Soil Water Content *59*
6. Water Movement in Soil *77*
7. Environmental Influence on Soil Formation *93*
8. Properties of Colloids in the Soil Environment *105*
9. Types of Soil Acidity *121*
10. Liming Acid Soils *131*
11. Soil Degradation by Salinity and Sodicity *143*
12. Soil Testing for Available Phosphorus *155*
13. Soil Organic Matter and Chemical Sorption *167*
14. Microbial Decomposition of Organic Materials in Soil *181*
15. Soil Survey Reports *197*

Appendix: Applications of Chemistry to Soil Science *207*

Preface

Our intention in this eighth edition is to build on the past strengths of this manual and continue to provide laboratory exercises illustrating important chemical, physical, and biological concepts about soil. The manual is intended to complement practically any textbook in an introductory soil science course. Users will find a balance, as expressed in the subtitle, between agricultural and environmental principles illustrated throughout the exercises.

Input from reviewers and our own experience in using the manual provided direction for this revision. Mostly minor refinements, as opposed to any sort of major revision, met these goals. Summarily we sought to accommodate users' suggestions by expanding discussions, clarifying procedural steps, adding uniformity to SI units, and bolstering the link between intended learning outcomes and student activities required in each lab.

Exercise 1, Soil as a Natural Resource, has a new Soil Profile and Site Evaluation Checklist and Soil Profile Characteristics reporting sheet. Exercise 3, Particle Size Distribution, now includes an additional procedure for evaluating the particle size distribution in a sand sample and matching it to United States Golf Association specifications for root zone mixes. Exercise 4, Bulk Density and Soil Porosity, includes a new figure on the volume composition of a surface soil, adds an equation for the calculation of percent solid space, and discusses paraffin as an alternative method for sealing the clod against water. Exercise 5, Soil Water Content, clarifies the distinction between a saturated soil and one at maximum water retention. Exercise 6, Water Movement in Soil, adds a section illustrating how soil texture and layering influence infiltration and capillary rise in soils. The title for Exercise 12 has changed to "Soil Testing for Available Phosphorus." Exercise 13, Soil Organic Matter and Chemical Sorption, addresses the pH dependent charge in soil humus and adds suggestions for the disposal of atrazine in soil.

The 15 exercises appear in many formats, including chemical analyses, suggested videos, field tour support, and physical measurements and evaluations. Students are required to use a mix of qualitative and quantitative assessments in completing the labs. Instructors are urged to adapt exercises by including soils appealing to local interests. However, soil selection is often critical to demonstrate optimum results and should be made with careful planning.

We express our gratitude to those involved in the legacy of this laboratory manual, and we aim to continue providing a useful resource for students learning about soil. We are indebted to our colleagues who have provided suggestions for improvement to this edition. We also wish to recognize Drs. Bill McFee and Jim Ahlrichs for their lasting impact on our education, about both soil science and soils teaching.

Steve J. Thien and John G. Graveel

Exercise 1: Soil as a Natural Resource

> *Soil is a vital natural resource for sustaining life and a quality environment on this planet. Conservation and appropriate management of soil relies on knowledge of both its form and function. This knowledge is difficult to acquire because soils are complex bodies that differ in many ways and for a wide variety of reasons.*
>
> *Exercise Goal: This exercise introduces useful concepts and practices for interpreting and evaluating soil properties. It can serve as a valuable reference for turf, agricultural, environmental, engineering, horticultural, landscape, and forestry applications where an understanding of soil is important. Field observations are described that can be used to interpret soil behavior, limitations, potentials, and management.*

Introduction to Soils

Soil is a vital resource for sustaining two basic human needs; a quality food supply and a livable environment. Along with air and water, soil contributes essential processes to the natural order of global cycles. This exercise introduces a study of soil form and function, the first step in linking the fundamental role of soil management and conservation to human survival on this planet.

Soils are a product of both inherited and acquired properties. Their current characteristics reflect an integration of original features with accumulated influences of subsequent environments. Deciphering important soil features requires keen senses of inquiry and observation. A goal of soil evaluation is to furnish knowledge useful for predicting soil behavior, limitations, and potential, all leading to appropriate use and management of this irreplaceable resource.

Soil refers to unconsolidated natural material on the earth's surface that supports plant growth. An individual soil is a three-dimensional body with recognizable boundaries. The interface with the atmosphere is the soil's upper boundary, the depth to which biological activity and weathering occur approximates the lower boundary, and laterally it is bounded by soil bodies with different properties. Because soils occupy definite positions on the landscape, individual soils can be mapped and named as a **soil series.**

Soil series are distinct, natural bodies differing in the composition and organization of their **profile,** a vertical cross section of layers, called **horizons** (Fig. 1-1). A landscape contains a continuum of individual soils, each distinguished by their different properties, processes, profiles, limitations, potential, and management requirements.

Soil characteristics result from the integrated effects of six **forming factors.** These factors are: parent material, topographic position, climate, organisms, time, and human activity. Because many variations of each factor are possible, a very large number of different soils can be identified. It's estimated that over 10,000 individually different soil series could be mapped in the United States alone.

Interpreting Observations

Every feature of a soil contributes some information toward understanding its nature. Untrained observers may notice only a few prominent features when first scrutinizing a soil, but evidence abounds for

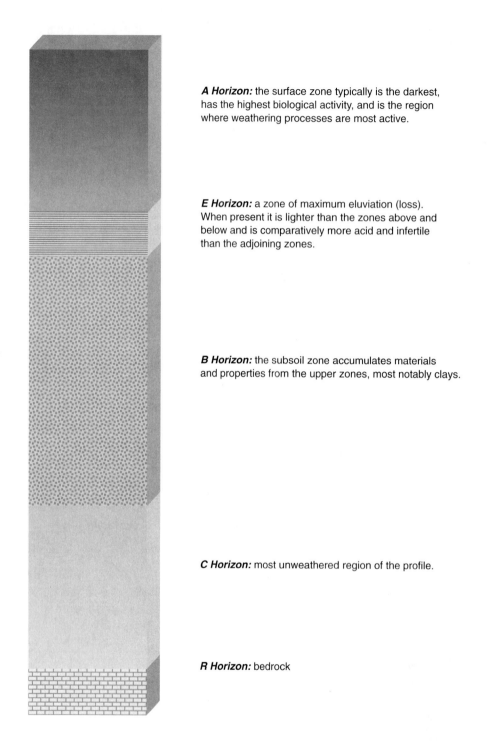

Figure 1-1. The soil profile is a vertical stack of distinguishable layers called horizons.

unlocking natural secrets. The trained eye recognizes color changes, particle distributions, cracking patterns, and many other variances as clues for explaining the past, present, and future nature of that soil. No discernible feature should be dismissed before establishing its influence.

Soil observations should be directed at both a property and a process level. A **soil property** is an identifiable, single quality or trait. Some examples would be: sand content, color, pH, slope, and porosity.

Some properties can be qualitatively estimated by simple observations in the field, while others require quantifiable laboratory analyses. This exercise acquaints you with 28 soil properties that can be assessed in the field.

A **soil process** identifies a series of actions causing a change in one or several properties. Erosion, deposition, organic matter decomposition, weathering, and acidification are examples of soil processes. (Table 1-1). Processes occur within soils and result in either additions or losses to the soil body, or in translocations or transformations within the soil body. Processes are usually assessed for their accumulated effects or intensity.

Table 1-1. Some processes occurring in soils.

Acidification	loss of basic cations and the gain of acidic cations causing pH to decrease
Aeration	replacement of soil air, usually O_2-depleted, with atmospheric air
Aggregation	formation of secondary units from binding sand, silt, and clay particles
Cementation	binding of soil material by $CaCO_3$ or oxides of iron, silicon, and aluminum
Decomposition	breakdown into constituent parts, mineral or organic
Deposition	material being laid down in a new place by agents like water, wind, ice, or humans
Differentiation	changes to soil material from the composite effects of soil development
Disintegration	breakdown of rocks and minerals into smaller particles by physical forces
Dissection	erosion topography where drainageways separate original landforms
Dissolution	solubilization of materials in water
Eluviation	removal of soil materials from a region within the soil profile by percolating water
Erosion	removal of the land surface by water or wind
Horizonation	evolution of distinct layers in soil deposits through natural actions
Illuviation	deposition into a lower horizon of soil material removed from an upper layer
Leaching	removal of soluble materials from soil
Littering	surface accumulation of organic material
Pedoturbation	mixing of soil by organisms, freeze/thaw, or wet/dry cycles
Salinization	accumulation of soluble salts in soil
Sodication	accumulation of exchangeable sodium in soil
Stratification	layering of soil materials by deposition processes
Sealing	dispersion and packing of the soil surface rendering it nearly impermeable to water
Weathering	physical and chemical changes caused primarily by atmospheric agents

Soil characterization is based on assessing the presence, absence, intensity, and interrelatedness of properties and processes. Knowledge of an individual soil property or process rarely provides sufficient information to guide proper soil use and management. Instead, a thorough **integrative evaluation** is needed in which all identifiable soil characteristics contribute information for assessing soil. This view presents soil as but one resource in a complex environment of many participants. The ecological adage that "everything affects everything else" rules out singular events and requires a holistic study of the nature of soil and surroundings. In a holistic study, soil interpretation and management must be based on an accurate, integrative evaluation of soil properties and processes. Grouping individual soils into **sequences** can assist the integrative evaluation by illustrating the impact of a prominent forming factor on related soils (Table 1-2).

Table 1-2. Concepts useful for an integrative evaluation of soil.

Anthroposequence	related soils that differ primarily because of the influence of humans
Biosequence	related soils that differ primarily because of the effect of plants, animals, and microorganisms
Chronosequence	related soils differing primarily as a result of time as a soil-forming factor
Climosequence	related soils differing primarily in properties associated with the effects of climate
Lithosequence	related soils differing primarily as a result of differences in parent material
Toposequence	related soils differing primarily because of topography's influence on soil formation
Land use	the nature and intensity of demands placed on the soil in relation to its natural limitations and potential
Soil quality	an integrated evaluation of the soil's ability to function for an intended use in accordance with its natural potential
Vegetative cover	an integrative, bioindicator of the soil's ability to support plant growth

Discovering Soil Differences

Look for differences in both the soil properties and processes that have occurred. When property and process differences are significant, soil behavior will change, as will the soil's response to management. While soil properties can be quantified by appropriate laboratory procedures, trained observation skills can often show relative differences and provide sufficient information to guide management decisions. This exercise illustrates field techniques for identifying soil differences and interpreting observations.

Using Soil Profiles

Materials:
- *Preserved or freshly collected soil profiles*
- *Munsell soil color book*
- *Soil survey report*
- *Soil maps*

Procedure:
- *Identify and rate all discernible differences among a set of soil profiles,*
- *suggest and develop an explanation about how observed differences affect soil use,*
- *interpret labeling on preserved profiles (if available),*
- *locate soil series on soil maps, and*
- *use soil survey report to furnish a complete profile description.*

Field Technique

Materials:
- *Munsell soil color book*
- *Soil survey report*
- *Soil thermometer*
- *Soil probe, auger, or shovel*
- *Dropper bottle with dilute acid*

- *Low power (2× to 5×) magnifying glass*
- *Water bottle*
- *pH test paper*

Procedure:
- *Select a representative field site, and*
- *observe soil by exposing a profile (or collect soil segments with an auger, probe, or spade and arrange segments to maintain depth orientation),*
- *look for clues, being as observant as a detective,*
- *identify and rate the properties listed in this exercise,*
- *complete the soil evaluation check list,*
- *delineate individual soils on a landscape on the basis of soil profile characteristics,*
- *describe how an individual soil can both occupy a unique niche and contribute to complex associations with other soils in a landscape,*
- *relate soil differences to soil behavior and potential response to management strategies, and*
- *maintain notes for future reference.*

How to Conduct a Field Evaluation

Soil can be evaluated at a field site by applying the definitions, evaluation methods, and expected outcomes supplied for 28 soil properties listed in this exercise. Evaluation relies on what you can see, feel, or smell, and in some cases, the use of common, inexpensive equipment. Like any talent, continued practice with the guidelines will increase mastery of your soil evaluation. Rank outcomes on the checklist at the end of the exercise and interpret their management implications.

People in turf, agriculture, environmental, engineering, horticulture, landscape, or forestry industries will find these guidelines useful. Novices can readily acquire an ability to identify and assess traits that will assist their communication about soil and their management of this vital natural resource.

You should be aware of some limitations to soil evaluation as done in this exercise. First, the evaluations presented here are largely qualitative in nature, designed to yield sufficient information to allow soil-to-soil comparison. Some soil properties, especially chemical or biological ones, cannot be reliably evaluated by field methods. For these instances and when more exact analysis of any soil property is required, quantitative laboratory tests should be used.

Each soil property listing includes the following information:

- **Definition**. *The property is defined in practical, simple-to-understand terms.*
- **Related Properties**. *A cross-reference list illustrates the interconnected nature of soil traits.*
- **Evaluation Method**. *Steps are given for field evaluation of the property.*
- **Outcomes**. *Expected results from using the evaluation methods are described.*
- **Implications**. *Properties and processes influenced by this property and relationships to soil use and soil management are listed to encourage and direct further evaluation.*

Categories of Soil Properties

Soil Physical Properties
Aeration
Aggregation (structure)
Color
Compaction (density, strength)
Cracks
Crusting
Penetration resistance
Permeability
Porosity
Temperature
Texture

Soil Organic and Chemical Properties
Odor
Organic matter
Organisms
pH
Rooting
Salinity
Sodicity
Thatch
Vegetative cover

Soil Water Relations
Dry spots
Infiltration
Water content
Water-holding capacity

Site Characteristics
Horizonation
Layering
Profile depth
Slope position

Guidelines for Evaluation of Soil Properties

AERATION: supplying air throughout the soil by exchange, or replacement, with air from the atmosphere. *(Related properties: color, compaction, porosity, rooting, sodicity)*

Evaluation method: Direct: use a hand lens to examine a soil core for amount and continuity of the largest pores. Indirect: (1) an abundance of fine, white roots suggests adequate aeration, or (2) red, brown, or tan soil colors confirm adequate aeration while gray or yellow colors denote poor aeration, or (3) well-aggregated soils provide ideal conditions for adequate aeration.

Outcomes: Rank zones within a soil core as well aerated, adequately aerated, or poorly aerated depending on the intensity of the indicators listed above.

Implications: Soil air composition, root and microbial respiration and growth, oxidation and reduction.

AGGREGATION (STRUCTURE): the arrangement of primary soil particles (sand, silt, and clay) into secondary units (aggregates) held intact by mineral and organic colloids acting as "glues." *(Related properties: aeration, compaction, organic matter, sodicity, texture)*

Evaluation method: Use gentle finger pressure to break intact soil into subunits. Observe the form and size of the aggregates that separate from the soil body.

Outcomes: Identify types as either: *Granular:* cubical, pea-size or smaller, dark, typically in surface soil; *Blocky:* cubical, larger than granular, with sharp edges (angular) or rounded edges (subangular), typically in subsoil; *Prismatic:* columnar shape, large or small, flat tops (prismatic) or rounded top (columnar), typically in subsoil; *Platy:* flat shape, large or small, typically in forested soils or compacted zones; *Single-grained:* typical of loose sands; or *Massive:* unaggregated, powdery or irregular sizes and shapes, zone is typically puddled or crusted, found in deep, unweathered layers or excessively tilled, compacted surface soils.

Implications: Root growth, aeration, water movement, water storage, infiltration, tilth, compaction, soil management, and microbial activity.

COLOR: a visual characteristic used for distinguishing one soil, or soil zone, from another. Three variables of color—hue, value, and chroma—allow for describing all soil colors. *(Related properties: aeration, organic matter, slope position, temperature)*

Evaluation method: Compare moist soil to chips in a Munsell soil color chart; use Munsell color notations found with profile descriptions in a soil survey manual; or generate a subjective name based on visual observation.

Outcomes: An example Munsell color notation is: 10YR 3/2 very dark gray, where 10YR = hue, 3 = value, 2 = chroma, and very dark gray is the color name. Subjective descriptions might list shades (very dark, dark, light, etc.) of black, brown, tan, red, yellow, gray, and white as either solid colors or mottles (spots).

Implications: Organic matter (humus) content, fertility, aggregation, tilth, microbial activity, aeration, drainage, water table characteristics, and soil oxygen status.

Guidelines for Evaluation of Soil Properties

COMPACTION (DENSITY, STRENGTH): an increase in bulk density and/or resistance to penetration (strength) as a result of some compressive force. *(Related properties: vegetative cover, infiltration, aeration, porosity, water content)*

Evaluation method: Look for resistance to probe penetration, poor vegetative cover, soil crusts lacking large pores, poor infiltration and/or ponded water, and zones with reduced root growth from poor aeration. Assess the site's compaction history in terms of amount and/or frequency of pressure, especially events that occurred when the soil was moist and highly susceptible to compaction.

Outcomes: Rate the apparent compaction as high, medium, or none based on the severity of the symptoms listed.

Implications: Aeration, root proliferation, infiltration, drainage, microbial activity, soil strength, residue decomposition, soil management, and nutrient availability. Increased compaction causes deterioration in soil physical, biological, and chemical properties.

CRACKS: temporary separation of the soil body because of drying; a break, slit, fissure, or opening. *(Related properties: infiltration, rooting, texture)*

Evaluation method: Cracks can be observed directly at the soil surface or where excavation has exposed the subsurface. Surface residue sometimes hides cracks. The extent of cracking can be explored with a wire or rated by water-filling capacity.

Outcomes: Rank cracking prevalence as none, little, moderate, or abundant. Describe the size of cracks as: fine (smaller than the width of a few pieces of paper), medium (between fine and large), or large (larger than the width of a pencil). Express cracking depth as shallow (at or near the surface only), moderately deep (throughout most of the surface horizon), or deep (extending into the subsurface).

Implications: Clay type, texture, wetting and drying cycles, infiltration, structure damage, and root damage.

CRUSTING: a condition where drying makes the surface layer more compact, hard, and brittle than the material immediately beneath it. *(Related properties: aeration, aggregation, cracks, infiltration, puddling)*

Evaluation method: Pick at the surface with a knife and observe if a hard, brittle layer exists. Crusts can be quite thin or very thick. Crusts are typified by irregular cracking patterns and an almost total lack of open pores (excepts for cracks). Crush pieces of crust in your hand to determine its strength (hardness).

Outcomes: Rank soils on whether or not they are crusted, the relative thickness of the crust, and the relative hardness.

Implications: Aeration, aggregation, cracks, porosity, seedling germination and emergence, infiltration, runoff, erosion, and soil management.

DRY SPOTS: irregular patches of dead or wilted plants where soils fail to wet despite adequate irrigation or precipitation. Also known as hydrophobic soil. *(Related properties: infiltration, permeability, water-holding capacity, vegetative cover)*

Evaluation method: Sprinkle area with water droplets. Normal soil has rapid adsorption of water, while on a hydrophobic soil the water beads for a considerable time. Note whether water beads form on thatch, soil, or both.

Outcomes: Rate soils as normal or hydrophobic.

Implications: Infiltration, wetting agents, water deficiency, and microbial activity.

Guidelines for Evaluation of Soil Properties

HORIZONATION: extent of the natural layering (horizons) in a soil profile. See Figure 1-2. *(Related properties: layering, profile depth, color, texture, aggregation)*

Evaluation method: Observe color, aggregation (structure), and/or textural changes from the soil surface downward. Assign a new horizon name whenever one of these properties change.

Outcomes: Designate master horizons present with the capital letter; O, A, E, B, C, or R. Two letters are used to name transition horizons (AE, AB, BC, etc.).

O Horizon: a zone of thatch, litter, or other organic accumulation on top of the first mineral layer.

A Horizon: the uppermost mineral layer having: accumulated humus, the darkest coloration, maximum biological activity, the most intense mineral weathering, and the greatest losses of soluble products by eluviation (downward movement with water). Usually the most fertile zone. Lack of A horizon denotes erosion or removal in land leveling.

E Horizon: a zone marked by extensive removal of clays and organic matter, lighter than layers above and below, usually acid and infertile. Common in acid soils in higher rainfall zones.

B Horizon: a zone of accumulated materials: clays, iron and aluminum oxides/hydroxides, humus, calcium carbonate, etc. High clay content increases water-holding capacity. Lack of a B horizon indicates a relatively young profile with limited weathering or developmental processes.

C Horizon: parent material little affected by formation processes and lacking definitive properties of horizons above.

R Horizon: bedrock.

Implications: Soil age, stage of maturity, and responsiveness to management.

INFILTRATION: the entry of water into soil. *(Related properties: texture, slope position, erosion, porosity, compaction, dry spots)*

Evaluation method: Low infiltration rate is associated with puddled or crusted sites, dry spots, clayey texture, eroded areas, and ponded sites. Low infiltration can be due to excessive runoff on steep slopes, but also look for a low amount of humus or loss of large pores at the soil's surface. Compare disappearance rate of free water.

Outcomes: Ratings of fast, medium, slow, and very slow should be adequate to separate soils into management categories.

Implications: Water runoff, erosion, aeration, soil texture, irrigation efficiency, residue management, soil conservation.

LAYERING: presence of stratified zones in a soil profile with different physical characteristics, primarily particle size. *(Related properties: porosity, texture, horizonation, drainage)*

Evaluation method: Examine a soil core for abrupt changes in soil texture, and perhaps porosity. Diffuse boundaries between layers present few water distribution problems. A low power magnifier can assist with this observation.

Outcomes: Describe the nature of layer changes, frequency of layers, and thickness of layers relative to their influence on soil behavior.

Implications: Human impact, deposition, landscaping, topdressing, soil amendments, water and root distribution, perched water table, drainage, and texture.

Guidelines for Evaluation of Soil Properties

ODOR: a soil property detectable by the sense of smell. *(Related properties: organic matter)*

Evaluation method: Smell a sample of freshly exposed soil. Avoid prolonged or vigorous inhalation.

Outcomes: Many odors can be detected in soils ranging in intensity from distinct to faint. Some typical odors are described below.

Implications: Volatile organic compounds, contamination, rotting or decomposing materials, microbial activities, pesticides, ammonia, and petroleum products.

ORGANIC MATTER (HUMUS): a relatively stable, dark-colored family of products resulting from the biological decomposition of organic materials. *(Related properties: color, organisms, aggregation, thatch, vegetative cover)*

Evaluation method: Humus content is proportional to dark soil coloration and generally decreases with depth. Less humus is required to darken coarse-textured soil than fine-textured soil. Humus particles cannot be seen with the naked eye.

Outcomes: Use a qualitative description of darkness (black, dark brown, light tan, etc.) or a Munsell color notation.

Implications: Soil color, aggregation, climate, native vegetation, water and nutrient adsorption, microbial activity, and carbon cycling.

ORGANISMS: the living component of soil. Includes flora (plants) and fauna (animals). Besides plant roots, the most important group is the microflora (bacteria, fungi, actinomycetes, and virus). *(Related properties: aggregation, dry spots, odor, organic matter)*

Evaluation method: Direct observation suffices for plant roots. Microflora must be indirectly detected by observing effects of their activity, such as: rate of residue decomposition, plant diseases, dry spots, aggregation, and soil darkening (humus). Detection of aromatic compounds indicates microbial activities.

Outcomes: Soils can be rated as being healthy with abundant, beneficial organismic activity; or as unhealthy, where little, or even harmful, activity is apparent.

Implications: Decomposition of organic residues, humus formation, soil aggregation, soil temperature, soil moisture, soil aeration, nutrient cycling, nutrient uptake, diseases, soil amendments, and root growth.

PENETRATION RESISTANCE: resistance to insertion of a rod into the soil. Also known as soil strength. *(Related properties: compaction, rooting, texture, water content)*

Evaluation method: Note the force required to push a bluntly pointed rod into the soil. Make comparisons under similar water contents and at similar depths.

Outcomes: Rank soil zones qualitatively as having a high, medium, or low resistance to penetration.

Implications: Root distribution, structural support, soil texture, soil moisture, compaction, and infiltration.

Guidelines for Evaluation of Soil Properties

PERMEABILITY (DRAINAGE): the ease with which air, water, or plant roots pass through the soil. *(Related properties: aeration, compaction, infiltration, layering, porosity, rooting)*

Evaluation method: Look for evidence that air, water, or roots from one region of the soil pass into another area. A uniform distribution of roots and water indicate good permeability. Sharp changes in water content or rooting between soil layers may signify a change in permeability. Look for aeration pores with a hand lens. Red, tan, or brown colors in the subsoil indicate adequate air permeability.

Outcomes: Rate permeability as low, moderate, or high.

Implications: Porosity, infiltration, drainage, layering, compaction, runoff, erosion, and distribution of air, water, and roots.

---

pH: a measure of the acidity or alkalinity of the soil environment. *(Related properties: organic matter, salinity, sodicity)*

Evaluation method: Place special pH papers in contact with wet soil and match their color to a pH chart.

Outcomes: Papers are available to distinguish between full pH units for common soil values.

Implications: All chemical and biochemical reactions have an optimal pH range. In conditions outside this range, reactions may be altered sufficiently to affect important processes related to plant growth. Optimum pH for most plant growth activities is slightly acid, near pH = 6.8.

POROSITY: that portion of the soil not occupied by solid material. *(Related properties: aeration, compaction, infiltration, structure, texture)*

Evaluation method: Use a low power magnifier to directly examine soil for abundance and size of pores. Red, tan, or brown soil colors and an abundance of roots are both good indicators of adequate porosity, especially large pores. Gray or yellow colors indicate low oxygen levels, perhaps caused by low porosity, or a low amount of large pores. Poor structure related to sodic conditions or compaction will indicate low porosity.

Outcomes: Describe the abundance (few, some, many), size (small, medium, large), and location of pores.

Implications: Root growth, microbial activity, water conduction and storage, aeration, compaction, soil texture, tillage, and aggregation.

PROFILE DEPTH: thickness of the soil down to the first layer that would impose some limitation to root growth of desired vegetation or other intended use. *(Related properties: horizonation, layering, soil temperature, rooting)*

Evaluation method: Examine the vertical dimension of a soil and assess the depth at which the first restrictive layer occurs. This may be bedrock, a strongly acid layer, an infertile layer, an impermeable layer, a dry zone, a saturated zone, a polluted zone, a zone with high salt content, etc.

Outcomes: Compare the useful depth of soil to the rooting depth of desired vegetation and indicate whether the profile depth is: deep (no limitations), moderate (conditions may limit profile depth), or too shallow.

Implications: Root growth, water and nutrient supply, soil temperature, microbial activities, landscape position, and buried layers.

ROOTING: abundance, location, and quality of plant roots. *(Related properties: aeration, compaction)*

Evaluation method: Examine a freshly exposed soil profile or extracted cores for the occurrence of healthy, white roots.

Outcomes: Express rooting with reference to quantity, depth, and healthiness of observed roots. For example: many healthy roots were found in the 0–6 inch layer, but few roots were located below 20 inches.

Implications: Plant growth, aeration, fertility, compaction, microbial activity, and erosion.

Guidelines for Evaluation of Soil Properties

SALINITY: the presence of sufficient dissolved salts in the soil solution to adversely affect plant growth. *(Related properties: water-holding capacity)*

Evaluation method: Saline soils have white, salty crusts wherever water evaporates. Plants on saline spots look dull, dark, and stunted. Plants often show leaf scorching, yellowing, and premature leaf drop.

Outcomes: Affected soils are termed saline.

Implications: Water absorption, plant growth, seed germination, drainage, irrigation, fertilization, and water quality.

SLOPE POSITION: the location an individual soil occupies in the landscape. See Figure 1-3. *(Related properties: erosion, horizonation, temperature, water content)*

Evaluation method: Compare location to diagram below.

Types:

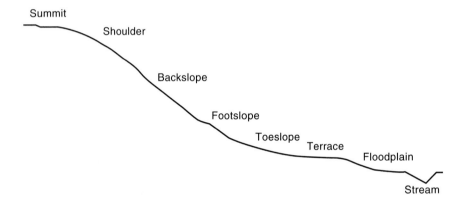

Implications: Soil management, erosion, creep, overland flow, rilling, deposition, textures, water storage, temperature, and plant growth.

SODICITY: the presence of sufficient exchangeable sodium on soil clays to adversely affect plant growth and soil structure. *(Related properties: aeration, infiltration, permeability)*

Evaluation method: Black crusts of dissolved organic matter appear in puddled spots. Soil aggregates are dispersed and surfaces appear sealed, resulting in low air and water permeability. Plants suffer from inadequate aeration.

Outcomes: Affected soils are called sodic. Other names describing the same condition are alkali and slick spots.

Implications: Aggregation, infiltration, waterlogging, aeration, irrigation, water quality, soil amendments, gypsum additions, and permeability.

Guidelines for Evaluation of Soil Properties

TEMPERATURE: the difference of solar heat gained by a soil body and the heat lost to surrounding bodies. *(Related properties: organisms, rooting)*

Evaluation method: Soil thermometer.

Outcomes: Typical values range from 0 – 35 °C (32 – 95 °F) in unfrozen soils, but seasonal, daily, and spatial variations can be large.

Implications: Plant growth, microbial activities, water content, slope, mulch, and soil color.

TEXTURE: classification of the mix of sand, silt, and clay in a soil. *(Related properties: most all soil properties carry some relationship to texture)*

Evaluation method: Moisten a ball of soil to the "putty" stage. Feed the soil between thumb and forefinger, flattening it into a ribbon. Allow ribbon to hang free as it grows. Note length of ribbon segments that break off on their own accord. The longer the ribbon segments, the greater the plasticity (clay content). Next, mix a pinch of soil and water together in your palm, noting degree of grittiness (sand) or smoothness (silt). Texture is also noted in the soil survey report.

Types: Assign texture class names from the following table.

	Ribbon Characteristics			
Dominant Wet Feel	No ribbon	Short ribbon, less than 2.5 cm (1 inch)	Medium ribbon	Long ribbon, greater than 5 cm (2 inches)
Smooth	Silt	Silt loam	Silty clay loam	Silty clay
Neither	Loam	Loam	Clay loam	Clay
Gritty	Sand	Sandy loam	Sandy clay loam	Sandy clay

Implications: Water storage, nutrient storage, pollutant retention, chemical activity, soil management, buffering capacity, water relations, and plant growth.

THATCH: the undecomposed or partially decomposed layer of organic matter above the mineral soil surface. When fully decomposed, this layer is called a mat. *(Related properties: dry spots, infiltration, layering, organic matter, organisms)*

Evaluation method: Look for the presence of organic debris on the soil's surface. If origin of the organic material is known, call it thatch. If decomposition precludes identification of the original material, call it mat. Check thatch for presence of insects, rodents, fungi, algae, or mold growth. Look for undissolved fertilizer or pesticide granules.

Outcomes: Rate amount of thatch (none, light, heavy) and identify other features noticed (molds, fertilizer granules, etc.).

Implications: Insect, rodent, microbial, and disease activity, water infiltration, nutrient and pesticide movement, rooting, aeration, and layering.

Guidelines for Evaluation of Soil Properties

VEGETATIVE COVER: the type, abundance, and quality of plants, especially in reference to soil type and site potential. *(Related properties: slope position)*

Evaluation method: Identify and map patterns of change in vegetation type, abundance, and quality, preferably on a soil map of the site. Other useful information that could be added to this map would include: recent amendments, response to treatment, and soil test data. Compare the most favorable ecological requirements (water, sunlight, nutrient, temperature, pH, rooting depth, etc.) of existing or desired species to actual conditions. Soil maps are available in soil survey reports, and favorable plant requirements are available in reference materials or from plant growth consultants.

Outcomes: List current vegetative types. Rank the abundance (sparse, adequate, or abundant) and quality (poor, medium, good) of existing vegetation.

Implications: Rooting, soil variability, slope, water availability, horizonation, pollution, fertility, and compaction.

WATER CONTENT: the proportion of a soil's water-holding capacity presently filled by water. *(Related properties: water-holding capacity, compaction, aeration, rooting, slope position, horizonation, layering, temperature, vegetative cover)*

Evaluation method: Squeeze a lump of soil in your hand and manipulate soil with fingers.

Excessively wet: soil will flow together and release liquid water. Soil with high clay content feels very sticky.

Wet: soil will be plastic (retains a molded shape) and feel damp to the skin. Plasticity increases with clay content.

Moist: soil will be friable (crumbly) and possess optimum properties for seedbed manipulation.

Moderately dry: soil begins to show cloddiness or dustiness; will wet cloth only after extended contact.

Dry: soil will either crumble, if sandy, or feel hard, if clayey; typically found in dusty or cloddy state.

Outcomes: Excessively wet, wet, moist, moderately dry, and dry. Note uniformity, or change, of water content between layers.

Implications: Compaction, aeration, rooting, slope position, layering, microbial activity, organic matter, soil color, temperature, soil texture, infiltration, and drainage.

WATER-HOLDING CAPACITY: the relative ability to retain water against the pull of gravity. *(Related properties: aeration, infiltration, permeability, water content)*

Evaluation method: Use the techniques described elsewhere to determine the texture and porosity of a soil. Evaluate these traits in all significant layers of the profile.

Outcomes: Rank soil water-holding capacity primarily on the amount of specific surface area (clay content) and secondarily on porosity. Plant-available water-holding capacity is usually of more interest than total water-holding capacity. "Available water" can be ranked according to the following scale:

Very low (1″ per foot or less):	sand
(2.5 cm per 30 cm)	loamy sand
	sandy loam, loam
Medium (about 2″ per foot):	silt loam and silt
(5 cm per 30 cm)	sandy clay loam
	clay loam
High (up to 3″ per foot):	silty clay loam and clay
(7.5 cm per 30 cm)	silty clay and sandy clay

Implications: Soil texture, porosity, compaction, aeration, infiltration, slope.

Soil as a Natural Resource

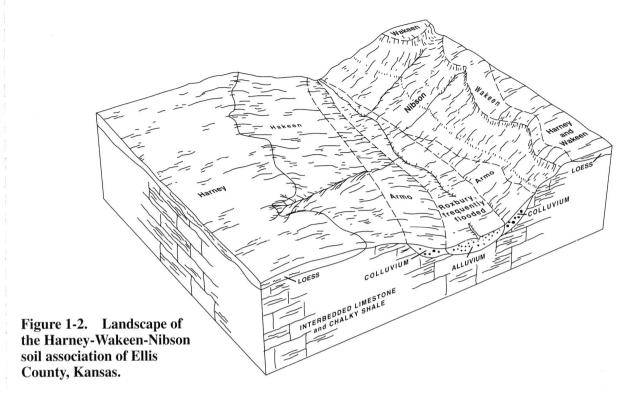

Figure 1-2. Landscape of the Harney-Wakeen-Nibson soil association of Ellis County, Kansas.

The Harney series consists of deep, well-drained soils formed in calcareous, medium-textured loess. These soils are nearly level to sloping and are found on upland sites.

The surface layer is typically 25 cm (10 inches) thick consisting of dark grayish-brown silt loam.

The subsoil is grayish-brown, about 75 cm (30 inches) thick, and has a silty clay loam texture. Permeability is moderately slow and available water capacity is high.

The underlying material is a light brownish-gray silt loam.

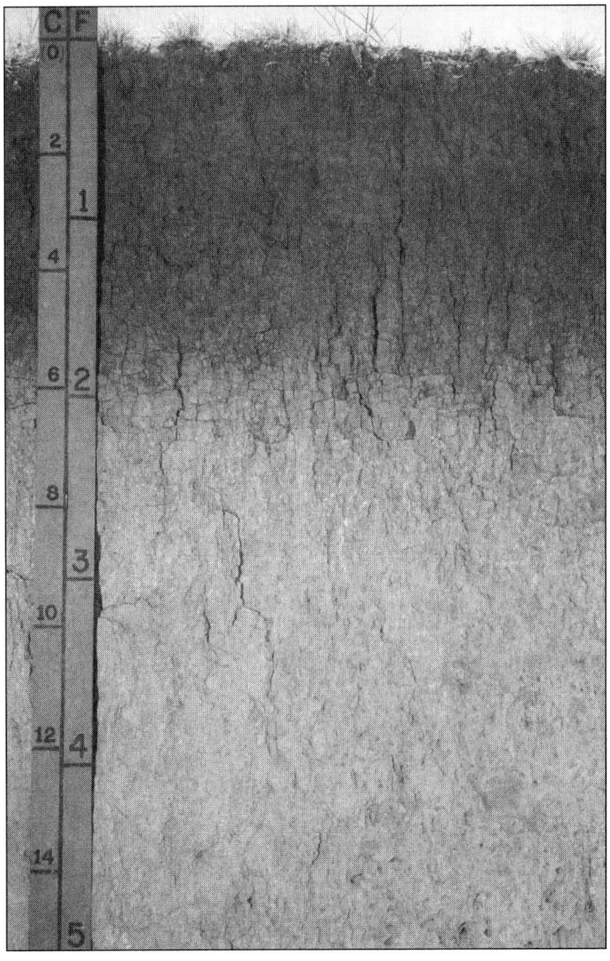

Figure 1-3. Harney silt loam profile. Complete profile descriptions are available at the National Soil Survey Center's website (http://www.statlab.iastate.edu/soils/nssc/).

15

Soil Profile and Site Evaluation Checklist:

Evaluate the properties listed using the criteria on the previous pages.

Soil Series: _____ Evaluator: _____

Physical Properties – A Horizon

AERATION: ☐ well ☐ adequate ☐ poor
AGGREGATION: ☐ granular ☐ blocky ☐ prismatic
☐ platy ☐ single grain ☐ massive
COLOR: ☐ v. dark ☐ dark ☐ medium ☐ light
COMPACTION: ☐ none ☐ medium ☐ high
CRACKS: ☐ none ☐ few ☐ moderate ☐ many
CRUSTING: ☐ none ☐ moderate ☐ severe
PERMEABILITY: ☐ high ☐ moderate ☐ low
POROSITY: ☐ many ☐ some ☐ few
SOIL STRENGTH: ☐ low ☐ medium ☐ high

Organic & Chemical Properties – A Horizon

ODOR: ☐ Type _____ ☐ none
☐ distinct ☐ moderate ☐ faint
ORGANIC MATTER: ☐ v. dark ☐ dark ☐ medium ☐ light
ORGANISMS: ☐ abundant ☐ few ☐ diseases
pH: ☐ Value: _____
ROOTING: ☐ many ☐ some ☐ few
SALINITY: ☐ saline ☐ nonsaline
SODICITY: ☐ sodic ☐ nonsodic
THATCH: ☐ none ☐ light ☐ heavy
PLANT COVER: ☐ abundant ☐ adequate ☐ sparse

Water Relations – A Horizon

DRY SPOTS: ☐ none ☐ few ☐ many
INFILTRATION: ☐ fast ☐ medium ☐ slow ☐ v. slow
WATER CONTENT: ☐ dry ☐ mod. dry ☐ moist ☐ wet
CAPACITY: ☐ high ☐ medium ☐ low

Mark horizon boundaries, depths, textures, and color notations on this blank profile. Use additional sheets for more horizons.

Physical Properties – B Horizon

AERATION: ☐ well ☐ adequate ☐ poor
AGGREGATION: ☐ granular ☐ blocky ☐ prismatic
☐ platy ☐ single grain ☐ massive
COLOR: ☐ v. dark ☐ dark ☐ medium ☐ light
COMPACTION: ☐ none ☐ medium ☐ high
CRACKS: ☐ none ☐ few ☐ moderate ☐ many
CRUSTING: ☐ none ☐ moderate ☐ severe
PERMEABILITY: ☐ high ☐ moderate ☐ low
POROSITY: ☐ many ☐ some ☐ few
SOIL STRENGTH: ☐ low ☐ medium ☐ high

Organic & Chemical Properties – B Horizon

ODOR: ☐ Type _____ ☐ none
☐ distinct ☐ moderate ☐ faint
ORGANIC MATTER: ☐ v. dark ☐ dark ☐ medium ☐ light
ORGANISMS: ☐ abundant ☐ few ☐ diseases
pH: ☐ Value: _____
ROOTING: ☐ many ☐ some ☐ few
SALINITY: ☐ saline ☐ nonsaline
SODICITY: ☐ sodic ☐ nonsodic
THATCH: ☐ none ☐ light ☐ heavy
PLANT COVER: ☐ abundant ☐ adequate ☐ sparse

Water Relations – B Horizon

DRY SPOTS: ☐ none ☐ few ☐ many
INFILTRATION: ☐ fast ☐ medium ☐ slow ☐ v. slow
WATER CONTENT: ☐ dry ☐ mod. dry ☐ moist ☐ wet
CAPACITY: ☐ high ☐ medium ☐ low

Site Characteristics & Soil-Forming Factors

Slope Position _____ Temperature: ☐ warm ☐ normal ☐ cool Profile Depth: ☐ deep ☐ moderate ☐ shallow

Climate _____ Native Vegetation _____ Parent Material _____ Topography _____ Time _____

Soil as a Natural Resource

Name _____

Soil Profile Characteristics for _____
(soil series name)

From either a soil survey manual or the soil series database at the National Soil Survey Center's website (http://www.statlab.iastate.edu/soils/nssc/), find information to describe the soil series seen on your field trip. Compare these professional descriptions to your observations recorded on the other side of this page.

Taxonomic class: _____

Major horizons:

Horizon	Depth	Texture	Color	Structure

Position in the landscape:

Drainage and permeability:

Principal use:

Primary distribution:

Soil Profile and Site Evaluation Checklist:

Evaluate the properties listed using the criteria on the previous pages.

Soil Series: _____ Evaluator: _____

Physical Properties – A Horizon
AERATION: ☐ well ☐ adequate ☐ poor
AGGREGATION: ☐ granular ☐ blocky ☐ prismatic
 ☐ platy ☐ single grain ☐ massive
COLOR: ☐ v. dark ☐ dark ☐ medium ☐ light
COMPACTION: ☐ none ☐ medium ☐ high
CRACKS: ☐ none ☐ few ☐ moderate ☐ many
CRUSTING: ☐ none ☐ moderate ☐ severe
PERMEABILITY: ☐ high ☐ moderate ☐ low
POROSITY: ☐ many ☐ some ☐ few
SOIL STRENGTH: ☐ low ☐ medium ☐ high

Organic & Chemical Properties – A Horizon
ODOR: Type _____ ☐ none
 ☐ distinct ☐ moderate ☐ faint
ORGANIC MATTER: ☐ v. dark ☐ dark ☐ medium ☐ light
ORGANISMS: ☐ abundant ☐ few ☐ diseases
pH: Value: _____
ROOTING: ☐ many ☐ some ☐ few
SALINITY: ☐ saline ☐ nonsaline
SODICITY: ☐ sodic ☐ nonsodic
THATCH: ☐ none ☐ light ☐ heavy
PLANT COVER: ☐ abundant ☐ adequate ☐ sparse

Water Relations – A Horizon
DRY SPOTS: ☐ none ☐ few ☐ many
INFILTRATION: ☐ fast ☐ medium ☐ slow ☐ v. slow
WATER CONTENT: ☐ dry ☐ mod. dry ☐ moist ☐ wet
CAPACITY: ☐ high ☐ medium ☐ low

Physical Properties – B Horizon
AERATION: ☐ well ☐ adequate ☐ poor
AGGREGATION: ☐ granular ☐ blocky ☐ prismatic
 ☐ platy ☐ single grain ☐ massive
COLOR: ☐ v. dark ☐ dark ☐ medium ☐ light
COMPACTION: ☐ none ☐ medium ☐ high
CRACKS: ☐ none ☐ few ☐ moderate ☐ many
CRUSTING: ☐ none ☐ moderate ☐ severe
PERMEABILITY: ☐ high ☐ moderate ☐ low
POROSITY: ☐ many ☐ some ☐ few
SOIL STRENGTH: ☐ low ☐ medium ☐ high

Organic & Chemical Properties – B Horizon
ODOR: Type _____ ☐ none
 ☐ distinct ☐ moderate ☐ faint
ORGANIC MATTER: ☐ v. dark ☐ dark ☐ medium ☐ light
ORGANISMS: ☐ abundant ☐ few ☐ diseases
pH: Value: _____
ROOTING: ☐ many ☐ some ☐ few
SALINITY: ☐ saline ☐ nonsaline
SODICITY: ☐ sodic ☐ nonsodic
THATCH: ☐ none ☐ light ☐ heavy
PLANT COVER: ☐ abundant ☐ adequate ☐ sparse

Water Relations – B Horizon
DRY SPOTS: ☐ none ☐ few ☐ many
INFILTRATION: ☐ fast ☐ medium ☐ slow ☐ v. slow
WATER CONTENT: ☐ dry ☐ mod. dry ☐ moist ☐ wet
CAPACITY: ☐ high ☐ medium ☐ low

Mark horizon boundaries, depths, textures, and color notations on this blank profile. Use additional sheets for more horizons.

Site Characteristics & Soil-Forming Factors

Slope Position _____ Temperature: ☐ warm ☐ normal ☐ cool Profile Depth: ☐ deep ☐ moderate ☐ shallow

Climate _____ Native Vegetation _____ Parent Material _____ Topography _____ Time _____

Soil as a Natural Resource

Name _____

Soil Profile Characteristics for _____
(soil series name)

From either a soil survey manual or the soil series database at the National Soil Survey Center's website (http://www.statlab.iastate.edu/soils/nssc/), find information to describe the soil series seen on your field trip. Compare these professional descriptions to your observations recorded on the other side of this page.

Taxonomic class: _____

Major horizons:

Horizon	Depth	Texture	Color	Structure

Position in the landscape:

Drainage and permeability:

Principal use:

Primary distribution:

Soil Profile and Site Evaluation Checklist:

Evaluate the properties listed using the criteria on the previous pages.

Soil Series: _____ Evaluator: _____

Physical Properties – A Horizon

AERATION: ☐ well ☐ adequate ☐ poor
AGGREGATION: ☐ granular ☐ blocky ☐ prismatic
☐ platy ☐ single grain ☐ massive
COLOR: ☐ v. dark ☐ dark ☐ medium ☐ light
COMPACTION: ☐ none ☐ medium ☐ high
CRACKS: ☐ none ☐ few ☐ moderate ☐ many
CRUSTING: ☐ none ☐ moderate ☐ severe
PERMEABILITY: ☐ high ☐ moderate ☐ low
POROSITY: ☐ many ☐ some ☐ few
SOIL STRENGTH: ☐ low ☐ medium ☐ high

Organic & Chemical Properties – A Horizon

ODOR: ☐ Type _____ ☐ none
☐ distinct ☐ moderate ☐ faint
ORGANIC MATTER: ☐ v. dark ☐ dark ☐ medium ☐ light
ORGANISMS: ☐ abundant ☐ few ☐ diseases
pH: Value: _____
ROOTING: ☐ many ☐ some ☐ few
SALINITY: ☐ saline ☐ nonsaline
SODICITY: ☐ sodic ☐ nonsodic
THATCH: ☐ none ☐ light ☐ heavy
PLANT COVER: ☐ abundant ☐ adequate ☐ sparse

Water Relations – A Horizon

DRY SPOTS: ☐ none ☐ few ☐ many
INFILTRATION: ☐ fast ☐ medium ☐ slow ☐ v. slow
WATER CONTENT: ☐ dry ☐ mod. dry ☐ moist ☐ wet
CAPACITY: ☐ high ☐ medium ☐ low

Physical Properties – B Horizon

AERATION: ☐ well ☐ adequate ☐ poor
AGGREGATION: ☐ granular ☐ blocky ☐ prismatic
☐ platy ☐ single grain ☐ massive
COLOR: ☐ v. dark ☐ dark ☐ medium ☐ light
COMPACTION: ☐ none ☐ medium ☐ high
CRACKS: ☐ none ☐ few ☐ moderate ☐ many
CRUSTING: ☐ none ☐ moderate ☐ severe
PERMEABILITY: ☐ high ☐ moderate ☐ low
POROSITY: ☐ many ☐ some ☐ few
SOIL STRENGTH: ☐ low ☐ medium ☐ high

Organic & Chemical Properties – B Horizon

ODOR: ☐ Type _____ ☐ none
☐ distinct ☐ moderate ☐ faint
ORGANIC MATTER: ☐ v. dark ☐ dark ☐ medium ☐ light
ORGANISMS: ☐ abundant ☐ few ☐ diseases
pH: Value: _____
ROOTING: ☐ many ☐ some ☐ few
SALINITY: ☐ saline ☐ nonsaline
SODICITY: ☐ sodic ☐ nonsodic
THATCH: ☐ none ☐ light ☐ heavy
PLANT COVER: ☐ abundant ☐ adequate ☐ sparse

Water Relations – B Horizon

DRY SPOTS: ☐ none ☐ few ☐ many
INFILTRATION: ☐ fast ☐ medium ☐ slow ☐ v. slow
WATER CONTENT: ☐ dry ☐ mod. dry ☐ moist ☐ wet
CAPACITY: ☐ high ☐ medium ☐ low

Mark horizon boundaries, depths, textures, and color notations on this blank profile. Use additional sheets for more horizons.

Site Characteristics & Soil-Forming Factors

Slope Position _____ Temperature: ☐ warm ☐ normal ☐ cool Profile Depth: ☐ deep ☐ moderate ☐ shallow

Climate _____ Native Vegetation _____ Parent Material _____ Topography _____ Time _____

Soil as a Natural Resource

Name _____

Soil Profile Characteristics for _____
(soil series name)

From either a soil survey manual or the soil series database at the National Soil Survey Center's website (http://www.statlab.iastate.edu/soils/nssc/), find information to describe the soil series seen on your field trip. Compare these professional descriptions to your observations recorded on the other side of this page.

Taxonomic class: _____

Major horizons:

Horizon	Depth	Texture	Color	Structure

Position in the landscape:

Drainage and permeability:

Principal use:

Primary distribution:

Exercise 2: Soil Texture

> *Soil texture characterizes the mineral fraction of soil by its proportions of sand, silt, and clay. Texture is considered a stable physical property of soil since the proportions of these particles change rather slowly. Because it's based on particle size, texture influences solid surface reactions, aeration, microbial, chemical, and water relations. Texture helps to determine the suitability of a soil for agricultural, engineering, and environmental use.*
>
> *Exercise Goal: This exercise describes the significance of soil texture as a physical property, illustrates use of the textural triangle for assigning soil textural class names, and demonstrates a method for determining soil texture without the use of laboratory equipment.*

Soil is a porous mixture of mineral particles, organic matter, air, and water with properties dependent on the nature and amount of each constituent. Mineral particles determine soil texture and typically comprise about one-half the volume of a soil in good condition for plant growth. A useful way to characterize the mineral fraction of soil consists of separating particles into three size categories, called **soil separates.** The three separates are; **sand** (2.0–0.05 mm), **silt** (0.05–0.002 mm) and **clay** (less than 0.002 mm). A **soil texture** name is assigned based on the proportions of sand, silt, and clay in a soil. The abundance or scarcity of the various separates affects nearly every aspect of soil use, behavior, and response to management.

Under natural conditions, particle size changes slowly, which makes soil texture a stable feature. Particle size (texture) within a soil profile initially reflects the distribution inherited from the parent material. Over time, weathering forms clays in the soil's surface that can then be removed by percolating water, a process called **eluviation.** Accumulation of these clays in lower parts of the profile, a process called **illuviation,** contributes to development of the B horizon. A soil's texture and, consequently, its behavior can be changed by adding finer or coarser materials. This is often done for greenhouse mixes, golf greens, or other special needs. Turf managers modify soil texture by coring and topdressing with sand. For most agricultural and forestry practices, it is not practical to change soil texture.

Perhaps the most important aspect of texture is the large surface area supplied by the many small particles. A small handful of soil may expose as much surface area as three or four football fields. As particle size decreases, the amount of surface area per unit weight, called **specific surface area,** increases proportionately. This means that when comparing equal masses, 0.1 mm particles will have 10 times more surface area than 1.0 mm particles. Sand exhibits very low specific surface area, and silt has only slightly more. However, both the fine state of subdivision of clay particles and the layered, platelike structure of clay minerals generates an abundance of external and internal surfaces. As a result, the clay fraction of a soil has a very large specific surface area.

Table 2-1 shows the distribution of surface area in three different soils. The clay fraction completely dominates the specific surface in a soil. For example, the silt loam has 20% sand and 20% clay, but the surface area contributed by the clay is 100,000 times that of the sand.

Table 2-1. Specific surface area of soil separates and the distribution of surface area in 100 grams of three soil textures.

Soil separate	Typical specific surface area	Sandy Loam	Silt loam	Silty clay loam
		Sand-silt-clay, %		
		65-25-10	20-60-20	15-55-30
	$cm^2\ g^{-1}$	— cm^2 —		
Sand	30	1,950	600	450
Silt	1,500	37,500	90,000	82,500
Clay	3,000,000	30,000,000	60,000,000	90,000,000

Mineral surfaces are important because they are the contact zone, or **interface,** for interaction between the particle and its surrounding environment. The amount of specific surface area in a soil governs the intensity of many important chemical reactions and physical processes. Because fine-textured soils contain high specific surface area and, therefore, an abundance of interfaces, they are more chemically and physically active than coarser soils. Soil features correlated to specific surface area include: soil management, plant nutrient adsorption, and pollutant buildup. Surfaces also provide sites involved in the formation of stable aggregates. These aggregates, in turn, determine soil porosity and regulate many aeration and water relations. Soil particles that expose large amounts of surface also weather more rapidly.

Knowledge of soil texture helps predict behavior of many other soil properties. A summary of how some water, tillage, erosion, chemical, and environmental properties relate to soil texture (Table 2-2) provides insight to this predictive relationship. For each situation listed, and others you can imagine, evaluate how the occurrence and abundance of mineral surfaces influences soil behavior. Knowing the correlation between soil composition and behavior is fundamental to understanding soils and using them appropriately. Soil property databases (survey manuals, websites, etc.) provide additional relationships between soil texture and suitability for building and construction use, urban planning, pollution control, and forest, range, or recreational use.

Table 2-2. Soil properties related to texture.

	Soil Texture Category		
	Coarse	**Medium**	**Fine**
Water relations			
Infiltration—entry of surface water into soil	Rapid	Medium/slow	Rapid, cracked; slow, not cracked
Percolation—water movement with the profile	Excessive	Good	Fair/poor
Water storage—plant-available water	Very low	Medium	High
Aeration—gas exchange with atmosphere	Very good	Moderate	Poor
Tillage			
Tillage power required—drawbar pull	Low	Medium	High
Tillability—ease of seedbed preparation	Easy	Moderate	Difficult
Consistence—degree of adhesion or cohesion			
Moist/wet	Loose	Friable	Sticky/plastic
Dry	Loose	Hard	Very hard/cloddy
Erosion			
Crusting potential	Very low*	Medium	High
Water erosion	Low	High	Low/moderate
Detachability—separation from other particles	High	Medium	Low
Transportability—removal from site if detached	Low	Medium	High
Wind erosion	High	Moderate	Low
Detachability	High	Medium	Very low
Transportability	Low	Medium	High
Agri-chemical relations			
Fertility potential—nutrient storage	Very low	Low	High
Buffering capacity—resistance to pH change	Very low	Low	High
Surface-active pesticide application rate	Low	Medium	High
Environmental considerations			
Leaching—solute transport	High	Medium	Low, if not cracked
Landfill site—suitability as base material	Poor	Poor	Good
Contaminant retention—adsorptivity	Low	Moderate	High
Air pollutant source	Low	Moderate/high	High

* Under some conditions very fine sands will crust.

Laboratory Activity

Part I. The textural triangle and its use

Soils can be grouped into one of 12 **soil textural classes** depending on their sand, silt, and clay content. These 12 classes and their composition limits are identified by using a **textural triangle** (Fig. 2-1). Soil class names are formed by using the terms *sand, silt, clay,* and *loam* as nouns, adjectives, or both. Note that *loam* is not a soil constituent, but a soil class consisting of mostly sand and silt with only a small amount of clay. **Loamy** is a lay term that typically refers to coarse-textured soils with a high amount of organic matter, but its use in reference to soil texture is not uniform.

The position of textural classes in the triangle is correlated to specific surface area. For soil classes on the same horizontal level, specific surface area increases from left to right. For soil classes on different tiers, specific surface area increases from bottom to top. Vertical positioning influences specific surface area more than horizontal placement because it reflects clay content.

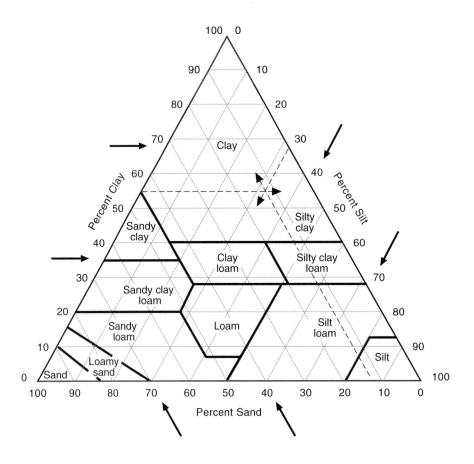

Figure 2-1. The textural triangle shows the percentages of sand, silt, and clay in the 12 textural classes. The intersection of the dotted lines shows a clay soil with a composition of 55% clay, 32% silt, and 13% sand. Note the direction used to plot data points from each of the axes.

Plotting composition data on a textural triangle to determine soil class name

If the composition of sand, silt, and clay is known, the data can be plotted on the textural triangle to determine the soil textural class name. All possible combinations of the three ingredients listed on the axes can be located within the boundaries of the triangle.

The procedure for plotting values on the triangle is as follows:

1. Locate the composition percentage on the appropriate axis and project a line parallel to the triangle's side that meets the zero point of that axis.

2. Repeat Step 1 for each soil separate. The soil class is found at the intersection of these projected lines.

Example: For a soil that contains 55% clay, 32% silt, and 13% sand, that procedure would be:

a. Locate 55% on the clay axis and project a line parallel to the bottom of the triangle in the manner shown by the dotted line in Figure 2-1.

b. Locate 32% on the silt axis and project a line parallel to the clay axis as shown in Figure 2-1. The intersection of these two lines occurs in the area designated "clay."

c. As a check, locate 13% on the sand axis and project a line parallel to the silt axis. If all three lines intersect at the same point, the class name has been determined correctly.

By definition, soil texture is determined using only the fine materials of less than 2.0 mm diameter. Thus, the sum of sand, silt, and clay composition must total 100%. Modifiers such as *gravelly, cobbly,* or *stony* describe material larger than 2.0 mm and are prefixed to textural names when their presence is significant.

3. Practice this plotting technique by completing Part I of the Data Sheet.

Part II. *Texture-by-feel: Determination of soil texture by plasticity and tactile analysis*

Soil texture can be identified simply by manipulating and feeling a moist sample. Clay content can be estimated by assessing the **plasticity** of moist soil, or its ability to be molded into, and retain, shapes when subjected to moderate pressure. Silt and sand content can be assessed by feeling (**tactile analysis**) a soil sample for the telltale smoothness of silt or the grittiness of sand. This procedure is useful whenever exact proportions are not required or when equipment necessary to obtain an exact analysis is not available, such as in the field. While proficiency with this method requires considerable practice, a novice can quickly obtain satisfactory results by learning a simplified textural triangle (Fig. 2-2) and applying the following technique.

A. The simplified textural triangle

1. Compare the modified textural triangle in Figure 2-2 to the triangle in Figure 2-1. The simplified version differs by omission of the typically well-known sand class and the rather uncommon silt class. The loamy sand class has been combined into the sandy loam category for simplicity and because both classes have similar properties. Boundaries for the remaining nine classes are symmetrically arranged to aid memory retention.

Soil Texture

2. Think of the simplified triangle as having three tiers based on clay content, or plasticity. The soils in the lowest tier, the **loam** group, have low clay content and exhibit little or no plasticity since they cannot be molded into ribbons, or form only short ribbons. The low specific surface area in these soils lends little or no **cohesiveness** (interparticle attraction) so molded shapes easily disintegrate.

The middle tier, the **clay loam** group, includes soils that are intermediate in clay content, cohesiveness, and ribbon formation. The upper tier, the **clay** group, includes soils with high clay content as evidenced by their high plasticity, high cohesiveness, and ability to be molded into long, stable ribbons.

3. The three tiers described in Step 2 can each be further subdivided into three columns based on a tactile analysis of sand and silt content. If the smooth feeling of silt is very prominent, the prefix *silt* or *silty* is added to the tier's group name, forming three classes on the right side of the modified triangle: silt loam, silty clay loam, or silty clay. If the gritty feeling of sand is very prominent, the prefix *sandy* is added to the tier's group name, forming three classes on the left side of the modified triangle: sandy loam, sandy clay loam, or sandy clay. If neither sand nor silt is judged predominant by the tactile analysis, the tier's group name is used without a prefix: loam, clay loam, or clay.

B. Determination of soil texture

A simple method exists for manipulating soil and identifying its soil textural class name. This method is called soil texture analysis by the "feel method." Use these instructions as you follow the flow diagram in Figure 2-3 to learn this manner of determining soil texture.

1. Moisten and knead a mass of soil about the size of a golf ball into a plastic, moldable condition.

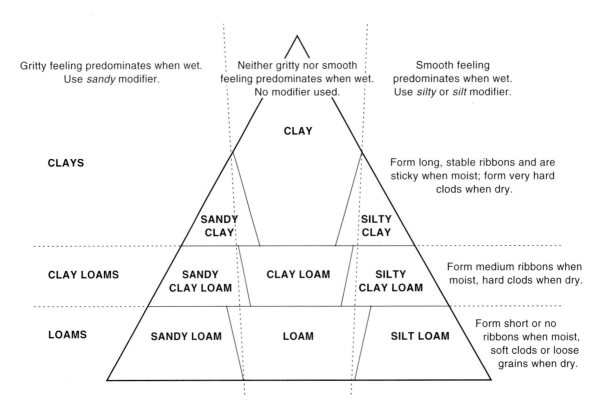

Figure 2-2. A simplified textural triangle helps you determine soil texture by the feel method.

2. Squeeze and feed soil between thumb and forefinger to form a ribbon of uniform thickness and width. Allow the ribbon to extend over the forefinger and break from its own weight. Continue ribboning the soil until several pieces have broken off. (This step can be difficult to master and may be aided by first rolling the soil into a cylinder of uniform diameter.) Ribbons should be jostled in the palm to gauge their strength and final length.

3. The length of the ribbon pieces depends on the clay content of the soil and delineates the tier of the simplified textural triangle to which the sample belongs.

- *Loams: samples form ribbons less than 2.5 cm (1 inch), suggesting less than 28% clay.*
- *Clay loams: samples form ribbons between 2.5–5 cm (1–2 inches), suggesting 28%– 40% clay.*
- *Clays: samples form ribbons longer than 5 cm (2 inches), suggesting greater than 40% clay.*

4. Next, the predominance of sand and silt in the sample is estimated by mixing water and a small bit of soil in the palm of your hand. Sand particles are large enough to feel gritty. Silt feels exceptionally smooth. If one of these feelings predominates, that prefix is added to the appropriate tier's group name. If neither grittiness nor smoothness predominates, then no prefix is added to the tier's group name.

5. With these two steps the texture of a soil can be identified by simple procedures requiring no laboratory equipment. Follow the flow diagram (Fig. 2-3) while studying the simplified textural triangle (Fig. 2-2) while you practice this technique on samples for which the texture is known. Then use the method to determine the texture of some unknown samples. Record your results on the data sheet.

6. Beware of the following considerations when using the texture-by-feel analysis:

- *Sands, especially the larger fractions, are easily overestimated because of their prominent feel.*
- *High amounts of organic matter can reduce the cohesion of clays and cause them to be underestimated.*
- *Clay types typically found in tropical and subtropical soils are not as plastic and cohesive as those types dominant in temperate region soils.*
- *Too much or too little water can alter the plasticity of a soil. Make sure you achieve expected results when working with samples of known texture.*

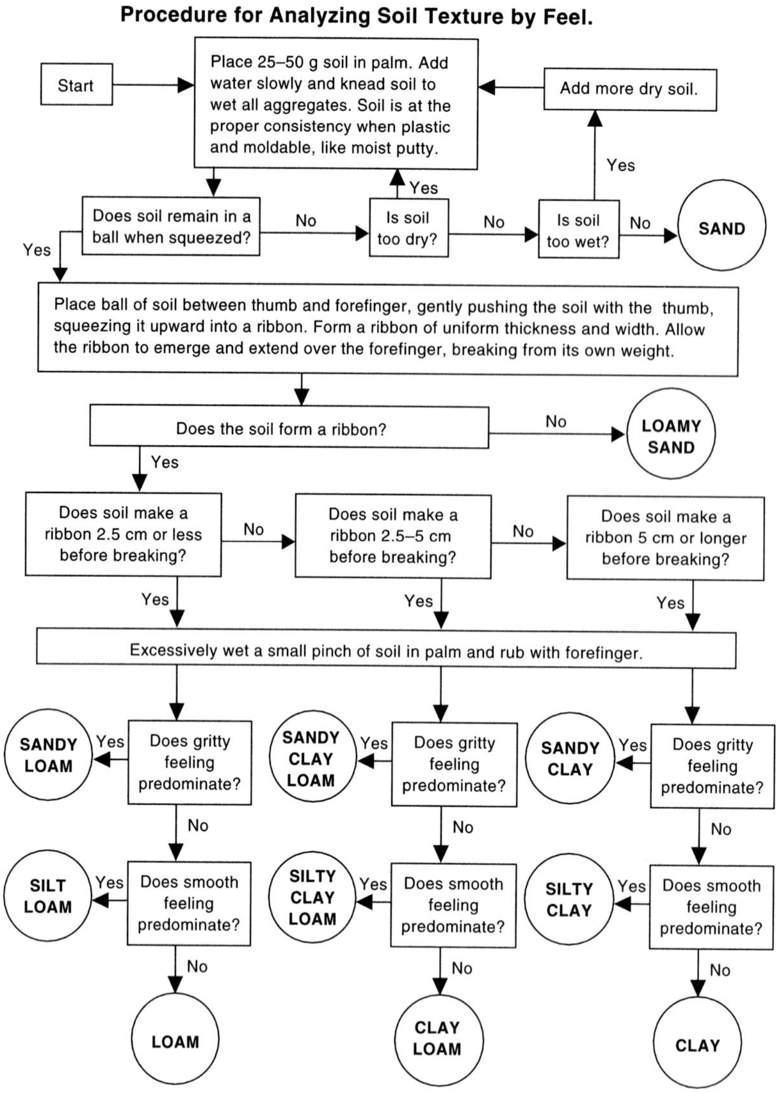

Soil Texture

Procedure for Analyzing Soil Texture by Feel.

[Flow chart: Start → Place 25–50 g soil in palm. Add water slowly and knead soil to wet all aggregates. Soil is at the proper consistency when plastic and moldable, like moist putty.

Does soil remain in a ball when squeezed? If No → Is soil too dry? If Yes, return to kneading step. If No → Is soil too wet? If Yes → Add more dry soil (return to kneading step). If No → **SAND**.

If Yes (ball holds): Place ball of soil between thumb and forefinger, gently pushing the soil with the thumb, squeezing it upward into a ribbon. Form a ribbon of uniform thickness and width. Allow the ribbon to emerge and extend over the forefinger, breaking from its own weight.

Does the soil form a ribbon? If No → **LOAMY SAND**.

If Yes:
- Does soil make a ribbon 2.5 cm or less before breaking? Yes →
- Does soil make a ribbon 2.5–5 cm before breaking? Yes →
- Does soil make a ribbon 5 cm or longer before breaking? Yes →

Excessively wet a small pinch of soil in palm and rub with forefinger.

Branch 1 (≤2.5 cm): Does gritty feeling predominate? Yes → **SANDY LOAM**. No → Does smooth feeling predominate? Yes → **SILT LOAM**. No → **LOAM**.

Branch 2 (2.5–5 cm): Does gritty feeling predominate? Yes → **SANDY CLAY LOAM**. No → Does smooth feeling predominate? Yes → **SILTY CLAY LOAM**. No → **CLAY LOAM**.

Branch 3 (≥5 cm): Does gritty feeling predominate? Yes → **SANDY CLAY**. No → Does smooth feeling predominate? Yes → **SILTY CLAY**. No → **CLAY**.]

Figure 2-3. Determining soil texture by the feel method. *Source:* "A Flow Diagram for Teaching Texture-by Feel Analysis," by Steve J. Thien, *Journal of Agronomic Education,* Vol. 8, 1979, pp. 54–55.

Soil Texture

Name _____ Section _____

Data

Part I. Determine the class name for these soils using the textural triangle in Figure 2-1

No.	Sand, %	Silt, %	Clay, %	Soil Class
1	33	33	34	Clay loam
2	10	68	22	
3	37	55	8	
4	52	8	40	
5	65	20	15	

Part II. Soil class names of unknown samples

Plasticity (long, medium, or short ribbon)	Predominant Wet Feel (smooth, gritty, neither)	Soil Class
1. _____	_____	_____
2. _____	_____	_____
3. _____	_____	_____
4. _____	_____	_____
5. _____	_____	_____

Soil Texture

Name _____ Section _____

Questions

1. What is the ratio of sizes for the largest sand, silt, and clay particle sizes? (Express this ratio so that the smallest unit receives a value of 1.) Name three everyday items that have a size ratio closely approximating that of sand, silt, and clay.

2. How can the processes of illuviation and eluviation affect soil texture?

3. Use data from Table 2-1 and calculate the specific surface area in the clay fraction of a 100 g soil sample containing 22% clay.

4. What two features of clay minerals account for their high specific surface area?

5. Select two examples from Table 2-2 and use specific surface area to explain why coarse- and fine-textured soils differ in behavior.

    a.

    b.

6. Define plasticity and explain why clay exhibits more plasticity than either sand or silt.

Soil Texture

Name _____ Section _____

7. Rank these six soil classes in order of ascending specific surface area. (1 = lowest, 6 = highest)

_____ Silt loam _____ Sandy clay loam _____ Sandy loam

_____ Silty clay _____ Silty clay loam _____ Loam

8. Summarize the relationship between soil texture and water erosion.

9. Explain how loam, sandy clay loam, and silty clay loam would respond to the texture analysis described in this exercise.

10. Red soils are commonly thought of as being clayey soils, and black soils high in organic matter are commonly referred to as loams. Does soil color determine soil texture? Does soil texture determine soil color? What relationships might exist between these two soil properties?

11. Explain why a water-soluble pollutant spill creates different hazards depending on the texture of the soil involved.

Exercise 3: Particle Size Distribution

> *The proportional distribution of sand, silt, and clay in a soil determines its textural class. A procedure for quantifying soil composition based on particle size distribution, called mechanical analysis, introduces the principles of dispersion and sedimentation rate and the use of a hydrometer. Application of Stoke's Law in the procedure provides the theoretical background for separating mineral particles by size based on their settling rate in suspension.*
>
> *Exercise Goal: This exercise requires an understanding of the principles of dispersion and sedimentation. Students must understand and apply Stoke's Law, prepare a soil suspension, and convert hydrometer measurements into a particle size distribution.*

The textural class of a soil is determined by its particle size distribution, namely, sand, silt, and clay content. Texture represents a rather stable soil characteristic and exerts an influence on many soil physical and chemical activities (see Exercise 2). This influence is directly related to the amount of surface activity presented by the mineral particles. Surface activity is a function of both particle size, which determines total specific surface area, and clay type, which determines relative surface reactivity. **Particle size distribution** analysis quantifies particle size categories, but does not determine clay type. Particle size distribution provides the information necessary for determining soil class on the textural triangle, an important standard for categorizing soil physical and chemical behavior on the basis of surface activity.

Quantitative determination of the proportions of differently sized solid particles is called **mechanical analysis**. This exercise uses the Bouyoucos hydrometer method of mechanical analysis, which relies on the principles of dispersion and sedimentation.

Dispersion: Individual soil particles must be **dispersed** (separated from each other) in an aqueous solution and remain dispersed to enable determination of particle size distribution (see Fig. 11-1). However, soils naturally exist as aggregates and not as a dispersed mixture of particles. **Aggregates** are secondary particles formed by cementing a mixture of **primary particles:** sand, silt, and clay. Cementing agents include organic matter, mineral oxides, or **polyvalent cations** (ions possessing more than one + charge). Dispersive methods remove or inactivate these binding agents. Only after binding forces have been negated can individual particles separate and their settling rate be properly analyzed.

Complete dispersion requires both mechanical and chemical assistance. Mechanical stirring overcomes weaker binding forces in large aggregates, but chemical agents are also necessary, especially to **deflocculate** clays. Polyvalent cations (normally Ca^{+2} and Al^{+3}) flocculate clays by forming interparticle, electrostatic links. Chemical dispersing agents (such as sodium hexametaphosphate) are effective in dispersing these clay bundles because:

- *The sodium monovalent cation (Na^+) replaces polyvalent cations adsorbed on clays, breaking the interparticle linkage. The displaced polyvalent cations form insoluble complexes with phosphorus, which prevents reestablishment of floccules.*
- *The adsorption of sodium, a highly hydrated cation, brings increased* **hydration** *of clays. This condition diminishes the binding strength between clay and cation, which raises clay particles' electronegativity and, hence, their repulsion from other clays.*

The mixture of dispersed soil particles in water is called a **suspension**. Once a true suspension state has been achieved, differential settling rates can be used to distinguish particle size distribution.

Sedimentation: The settling rate of a mineral particle in water, **sedimentation**, depends on the size of the particle. Large particles settle out of suspension more rapidly than small particles do. Analytical techniques based on this direct sedimentation relationship allow quantification of particle size distribution.

The connection between particle size and settling rate is expressed by **Stoke's Law**. This relationship shows that small particles, those exposing high specific surface area (m² g⁻¹), produce more resistance to settling through the surrounding solution than do large particles and, hence, settle at slower velocities.

$$\text{Stoke's Law} \quad V = \frac{D^2 g (d_1 - d_2)}{18n}$$

The formula shows that the settling velocity, V, is directly proportional to the square of the particle's effective diameter, D; the acceleration of gravity, g; and the difference between the density of the particle, d_1, and density of the liquid, d_2; but inversely proportional to the viscosity (resistance to flow) of the liquid, n. The density of water and its viscosity both change in a manner so that particles settle faster with increased temperature. Hence, it may be necessary to apply temperature correction factors as explained within the procedure.

Stoke's Law can be condensed to $V = kD^2$ by assuming constant values for all components except the effective diameter of soil particles. Then, for conditions at 30° C, $k = 11241$. For particle size values in centimeters, the formula yields settling velocity, V, in centimeters per second. Because soil particles do not meet the requirement of being smooth spheres, exact conformance to Stoke's Law is not realized.

Soil erosion into surface waters provides an environmental application of sedimentation principles and illustrates another criteria of Stoke's Law, namely, that the settling solution should be still. Moving water maintains particles in suspension that would otherwise settle in still water. Thus, sediment loads in streams and rivers are determined by water velocity and turbulence. Sediments segregate by particle size during settling. Fast-moving water can transport even very large particles, but as water flow slows, first sand particles and then silt are deposited. These deposits can bury an existing surface, alter subsequent water flow patterns, and reduce reservoir capacity. Clays, on the other hand, settle only when water movement has virtually ceased. Since clays constitute the most surface active fraction and bind chemicals that can constitute pollutants, clay deposits are frequently sites of environmental pollution.

This procedure uses a hydrometer to quantify the solid material remaining in suspension at each stage of the sedimentation process. The **hydrometer** is calibrated to measure the density of a suspension at the hydrometer's center of buoyancy in units of grams per liter. Research has determined that within 40 seconds sand particles (0.05 mm and larger) have settled below the buoyancy center of the hydrometer. Within two hours, silt particles (0.05–0.002 mm) have similarly settled and no longer influence the hydrometer. Thus, measuring the density of the soil suspension 40 seconds after shaking and again at 2 hours provides the information necessary to calculate the percentages of sand, silt, and clay in a soil.

Particle Size Analysis of Sand

The United States Golf Association (USGA) Green Section has established specifications for the construction of putting greens on golf courses. The success of USGA greens depends largely on having the proper physical characteristics in the root zone mixture. Standards for infiltration, percolation, porosity, density, and water retention capacity have been established, but native soils that meet these specifications are almost nonexistent. Therefore, most root zone mixes are prepared using specific-sized sand (80–90%) and organic amendments (10–20%).

With the sand fraction having such a large impact on the physical properties of the root zone, the USGA Green Section has specified the sand particle sizes that meet these specifications. In this exercise you will test a sand sample and determine whether it complies with the requirements shown in Table 3-1.

Table 3-1. USGA sand specifications for golf green root zone mixes.

Particle size (mm)	Name	Ideal content	Recommended (by weight)
3.4–2.0	Fine gravel	None	Not more than 10%, including a maximum of 3% fine gravel.
2.0–1.0	Very coarse sand	None	
1.0–0.5	Coarse sand	100%	Minimum of 60% of the particles must fall in this range.
0.5–0.25	Medium sand		
0.25–0.1	Fine sand	None	Not more than 20%.
0.1–0.05	Very fine sand	None	Not more than 5%.
<0.05	Silt & clay	None	Not more than 5%.

Particle Size Distribution

Laboratory Activity

Part I. Bouyoucos hydrometer analysis

Materials

- Soil hydrometer, ASTM No. 152H with Bouyoucos scale in grams per liter.
- Sedimentation cylinder, with 1000 mL mark
- Dispersing solution: dissolve 35.7 g technical grade sodium hexametaphosphate, $(NaPO_3)_6$, and 7.9 g sodium carbonate, Na_2CO_3, in about 900 mL deionized water. Adjust pH to 8.3 with additional sodium carbonate. Bring final volume to 1000 mL.
- Thermometer
- Balance
- Mechanical mixer with stirring cup
- 30% hydrogen peroxide, H_2O_2, for optional procedure only.
- Sieve, 300 mesh (50 μm openings), for optional procedure only.

Procedure

1. Weigh 40 g oven-dried soil into a stirring cup. Fill the cup half full with deionized water and add 10 mL of dispersing solution. Mix and let stand for 10 minutes to initiate chemical dispersion.

2. Place cup on mixer and stir 3 minutes for coarse-textured soils and 4–5 minutes for high clay soils. This step completes mechanical and chemical dispersion.

3. Quantitatively transfer stirred mixture to a sedimentation cylinder with a stream of deionized water. Fill cylinder to the 1000 mL mark with deionized water.

4. Agitate mixture to *uniformly suspend all* material throughout the liquid. This step can be done by one of two methods, either

   a. stopper and invert the cylinder several times (but do not shake in a manner that would produce circular currents in the liquid as this can alter the settling rate), or

   b. insert a plunger (circular plate on the end of a rod) and move it up and down until all material is equally distributed.

5. When suspension is complete, set cylinder on table (or remove plunger) and mark the time. After about 30 seconds, slowly lower the hydrometer into the suspension and release it. Then read the scale at the meniscus exactly 40 seconds after agitation was stopped. Remove hydrometer. Repeat steps 4 and 5 until hydrometer readings within 0.5 g of each other are obtained and record that reading on the data sheet.

6. Hang a Celsius thermometer in the suspension for about 3 minutes, record the temperature, and calculate the corrected hydrometer reading.

   - For each degree above 18 °C, add 0.25 g/L to the original hydrometer reading.
   - For each degree below 18 °C, subtract 0.25 g/L from the original hydrometer reading.

7. Mix the suspension again as described in Step 4 and place the cylinder where it will not be disturbed. After 2 hours, insert the hydrometer into the suspension and take another reading. Record the temperature and correct this hydrometer reading as described in Step 6.

8. Discard the liquid portion from the sedimentation cylinder into sink drains. Transfer settled materials to a designated waste receptacle.

Part II.   Optional procedure: Increased precision of sand and clay analysis

The procedure described in Part I typically overestimates sand content by 6%–10% because convection currents in the cylinder generated by achieving full suspension reduce their settling rate. This alternate procedure uses sieves to separate sand from the finer mineral fractions. These sieves are costly and the screening is easily damaged. Appropriate care is needed to preserve their integrity.

Other ways that this optional procedure increases precision include using a hydrogen peroxide digestion to eliminate organic materials that might prevent complete soil dispersion. Also, while a 2-hour settling time yields satisfactory results for many applications, some very fine silt may be included in the clay reading. An 8-hour settling time is used here and will increase the precision of clay content analysis.

1. Change Step 1 as follows: weigh soil into a beaker, add 30 mL water, and stir. Cautiously add a few mL of 30% H_2O_2 (hydrogen peroxide) and stir. Each time reaction subsides, add additional H_2O_2 in the same manner until no additional reaction is observed. Complete the removal of organic matter by heating beaker for 1 hour at no more than 90 °C. Quantitatively transfer mixture from beaker to stirring cup and proceed as indicated.

2. Alter the remainder of the procedure by omitting Step 5, increasing the settling period in Step 7 to 8 hours, and omitting Step 8.

3. After Step 7, wash all soil materials from the cylinder with a stream of water onto a 300 mesh sieve. Wash all silt and clay through the sieve with a gentle agitating stream of water administered from below the sieve's screen. Avoid contacting the sieve screen with any object as this material is easily damaged. Discard material washed through the sieve.

4. After the sample is washed free of silt and clay, transfer the sand to a beaker with a stream of water. Allow the sand to settle, then decant the excess water from the sample. Dry the sand to a constant weight using microwave energy (approximately 20 minutes) or a hot air oven at 105 °C (approximately 24 hours).

5. Weigh dried sand and beaker. Remove sand and weigh empty beaker.

Part III.   Procedure for particle size analysis of sand

1. Prepare the stack of sieves by arranging the appropriate sieves (see Table 3-2) into column. Start with the bottom pan and add sieves with progressively larger openings.

Table 3-2. Sieve number and openings size.

Opening mm	Opening inches	US Standard Sieve No.
2.00	0.08	10
1.00	0.04	18
0.50	0.02	35
0.25	0.01	60
0.105	0.004	140
0.053	0.002	270

Particle Size Distribution

2. Add a 50 g sample of dry sand into the top sieve, place lid onto sieve column, and shake.

3. Transfer the sand retained on each sieve into a preweighed plastic dish and weigh sand plus dish. Subtract weight of the dish and record the values on the data table.

DO NOT RUB THE SCREEN WITH ANY OBJECT.
USE ONLY A SOFT BRUSH TO DISLODGE PARTICLES CAUGHT IN THE SCREENS.

Name(s) _____ Section _____

Data

I. Particle size distribution—standard precision

1. Soil identification number or name _____

2. Soil weight, g _____

3. Average 40-second hydrometer value, g/L _____

4. Temperature of suspension @ 40-sec, °C _____

5. Temperature-corrected 40-sec hydrometer value, g/L _____

6. Two-hour hydrometer value, g/L _____

7. Temperature of suspension @ 2 hour reading, °C _____

8. Temperature-corrected 2-hour hydrometer reading, g/L _____

9. Grams of sand (Line 2 − Line 5) _____

10. Grams of clay (Line 8) _____

11. Grams of silt (Line 2 − Line 9 − Line 10) _____

12. Percent sand (Line 9 ÷ Line 2) × 100 _____

13. Percent clay (Line 10 ÷ Line 2) × 100 _____

14. Percent silt (Line 11 ÷ Line 2) × 100 _____

15. Soil textural class _____

Particle Size Distribution

Name(s) _____ Section _____

Data

II. Particle size distribution—high precision

1. Soil identification number or name _____

2. Soil weight, g _____

3. Weight of beaker and dried sand, g _____

4. Weight of beaker, g _____

5. Weight of sand, g (Line 3 − Line 4) _____

6. Eight-hour hydrometer value, g/L _____

7. Temperature of suspension @ 8-hour reading, °C _____

8. Temperature-corrected 8-hour hydrometer reading, g/L _____

9. Grams of sand (Line 3 − Line 4) _____

10. Grams of clay (Line 8) _____

11. Grams of silt (Line 2 − Line 9 − Line 10) _____

12. Percent sand (Line 9 ÷ Line 2) × 100 _____

13. Percent clay (Line 10 ÷ Line 2) × 100 _____

14. Percent silt (Line 11 ÷ Line 2) × 100 _____

15. Soil textural class _____

Particle Size Distribution

Name(s) _____ Section _____

Data

III. Particle size analysis of sand

Particle	Size mm	Sieve No.	Sample 1 g	Sample 1 %	Sample 2 g	Sample 2 %
Composite sample	mixed	n.a.		100%		100%
Fine gravel	3.4–2.0	on 10				
Very coarse sand	2.0–1.0	on 18				
Coarse sand	1.0–0.5	on 35				
Medium sand	0.5–0.25	on 60				
Fine sand	0.25–0.1	on 140				
Very fine sand	0.1–0.05	on 270				
Silt and clay	<0.05	thru 270				
Compliance with USGA standards for root zone mixes? (circle answer)				YES NO		YES NO

43

Particle Size Distribution

Name(s) _____   Section _____

Questions

1. Use the condensed version of Stoke's Law ($V = 11241D^2$) to calculate the following. (Note: Values for D must be in centimeters and the resulting value of V will be in centimeters per second). How long would it take the largest medium sand particles to settle to the bottom of a quiet pond that is 2 meters deep? How long would it take the smallest silt-sized particles?

2. Why are sandy deposits found near the banks of a flooding river, while silt-laden sediments are found farther away from the river channel?

3. Explain how a hydrometer works.

4. Suppose that all the silt and clay was transferred from the mixing cup to the settling cylinder, but that some of the sand remained in the cup. How would this affect your calculated values for sand, silt, and clay percentages? Explain.

5. Describe the difference between dispersed and flocculated colloids, referring to how they might appear in nature and how their chemistry would differ.

Particle Size Distribution

Name(s) _____ Section _____

6. Describe a natural illustration of the principles of sedimentation studied in this exercise.

Questions

7. Erosion can cause soil dispersion and form separate deposits of sands, silts, and clays. Why do each of these deposits form unique environmental hazards?

8. Why are there particle size criteria for golf green root zone mixes?

Exercise 4: Bulk Density and Soil Porosity

> *The pore system in soils facilitates air exchange, water distribution and storage, root growth, and microbial activities. Total pore space and the distribution of macro-, meso-, and micropores regulate important soil processes related to plant growth and environmental quality. Porosity is primarily governed by soil texture, aggregation, and compaction, properties collectively measured as bulk density. Generally, low bulk density (high porosity) is preferred in agricultural applications, and high bulk density may be most useful in engineering situations.*
>
> *Exercise Goal: You will learn how to determine soil porosity by measuring bulk density and using a standard particle density value. This exercise illustrates several methods of sampling soils for bulk density analysis. Archimedes' Principle is presented for determining the volume of an irregular body and calculating bulk density and porosity values.*

Soil is a unique arrangement of solids, liquids, and gases. The liquid phase (**soil solution**) and the gas phase (**soil air**) occupy voids in the solid phase, called **pores.** The pore system in soil provides the conduits for air and water exchange with its environment and houses root and microbial activities. **Soil porosity,** the amount of pore volume, provides a useful indicator of soil behavior. For example, the balance and composition of liquid and gas constituents in the pores determine the suitability of a soil for plant growth. A medium-textured, well-aggregated soil will contain about 50% pore space and be in good condition for plant growth when the pores hold an equal distribution of nutrient-rich water and well-oxygenated air (Fig. 4-1).

Because pore size affects pore activity, many important soil processes are closely dependent on pore size distribution. **Macropores** facilitate free-water drainage, aeration, evaporation, and gas exchange; **mesopores** are essential to capillary water distribution; and **micropores** provide water storage sites. Macropores are most prevalent in sandy soils and well-aggregated soils, but can be converted to micropores by compaction. Medium-textured soils have an abundance of mesopores. Clays promote

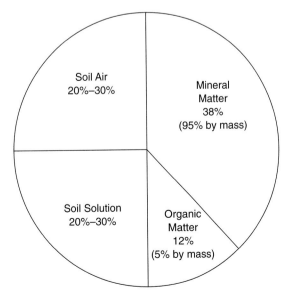

Figure 4-1. Relative volume of the four principal components of a medium-textured, well-aggregated soil.

aggregation, but can also be readily compacted. Clays also increase water storage by providing an abundance of micropores. Thus, texture and structure, plus the level of induced compaction, are the main properties governing amount and type of pore space in soil. Organic matter affects porosity through its enhancement of soil aggregation.

Porosity and pore size distribution influence many important processes that impact plant growth and environmental quality. These include: aeration, drainage, infiltration, erosion, runoff, root distribution, water storage, and nutrient availability. Soil management that alters pore properties in soil can have unintended consequences.

Porosity, or percentage of pore space, can be calculated if bulk density and particle density are known. The **bulk density** of a soil is its mass (weight) per unit volume. In its natural state, a soil's volume includes solids and pores; therefore, a sample must be taken without compaction or crumbling to correctly determine bulk density. A suitable sample can be obtained by any method that forces a straight-sided container of known volume into the soil without altering the area being sampled. The container is then dug out and trimmed of excess soil, whereupon the soil in the sampler can be removed, dried to a constant weight, and weighed. The bulk density is calculated by finding the ratio of the sample's weight (g) to its volume (cm^3), as follows:

$$Bulk\ density = \frac{Oven\ dry\ soil\ weight,\ g}{Volume\ of\ soil\ solids\ and\ pores,\ cm^3} \qquad Eq.\ 4\text{--}1$$

The bulk density of mineral soils commonly ranges from 1.1 to 1.5 g cm^{-3} in surface horizons. It usually increases with depth and tends to be high in sands and compacted or "pan" horizons and low in soils with abundant organic matter. Tillage operations loosen soils and temporarily lower bulk density, while compaction processes raise bulk density. High bulk densities correspond to low porosity. Natural soil-forming processes that increase aggregation reduce bulk density, but excessive tillage and raindrop impact on bare soil destroy aggregation and increase bulk density.

Particle density expresses the volumetric mass of the soil's solid phase. Particle density is the weight per unit volume of the soil solids, differing from bulk density because the volume used in this instance is only that of the solids and does not include the pore spaces. Particle density can be calculated as follows:

$$Particle\ density = \frac{Oven\ dry\ soil\ weight,\ g}{Volume\ of\ soil\ solids,\ cm^3} \qquad Eq.\ 4\text{--}2$$

Particle density represents the average density of all the minerals composing the soil. For most soils this value is very near 2.65 g cm^{-3} because quartz has a density of 2.65 g cm^{-3} and quartz is usually the dominant mineral. Use 2.65 g cm^{-3} as the particle density value in porosity calculations unless directed otherwise. Particle density varies little between soils and has little practical significance except in the calculation of pore space.

Porosity is that portion of the soil volume occupied by pore spaces. This property does not have to be measured directly since it can be calculated using values determined for bulk density and particle density. Finding the ratio of bulk density to particle density and multiplying it by 100 calculates the percent solid space, so subtracting it from 100 obviously gives the percent of soil volume that is pore space.

$$\text{Solid space, \%} = \left(\frac{\text{Bulk density}}{\text{Particle density}} \times 100\right) \qquad \text{Eq. 4--3}$$

$$\text{Porosity, \%} = 100 - \left(\frac{\text{Bulk density}}{\text{Particle density}} \times 100\right) \qquad \text{Eq. 4--4}$$

Sample Calculation of Porosity

A 260 cm³ cylindrical container was used to collect an undisturbed soil sample. The container and soil weighed 413 g when dried. When empty, the container weighed 75 g. What is the bulk density and porosity of the soil?

A. To determine bulk density:

Sample volume = 260 cm³

Sample weight = 413 − 75 = 338 g

Bulk density = $\frac{338\ g}{260\ cm^3}$ = 1.3 g cm^{-3}

B. To determine porosity:

Bulk density = 1.3 g cm^{-3}

Particle density = 2.65 g cm^{-3}

Porosity = $100 - \left(\frac{1.3}{2.65} \times 100\right)$ = 51%

Example Problem Set

1. What is the bulk density of a soil sample that weighs 360 g when dry and occupies 300 cm³? What is the porosity of this sample?

2. Bulk density samples were collected from a wheel-tracked zone and from adjacent uncompacted soil. How do the bulk density and porosity differ at these two sites?

Wheel-tracked site		**Uncompacted site**
610	Sample volume, cm³	545
1300	Weight of container and dry soil, g	900
250	Weight of container, g	170

3. Dig deeper! A cylindrical soil core taken from the field measured 10 cm high and 6 cm in diameter and weighed 720 g in a field-moist condition and 500 g when dried to a constant weight. What is the bulk density of this sample? What is the porosity for this sample?

Laboratory Activity

Sampling Procedures

A good time to collect bulk density samples is during a field trip. Comparisons can be made between soils from recently tilled fields and from pastures, forest, and fields in crops ready to be harvested; between surface soils and subsoils; between soils in wheel tracks of heavy equipment and nearby uncompacted soil; between soils from student paths on campus and from nearby untrampled grassy areas; between crusted and aggregated sites; and so on. Prepare a composite table based on the samples collected by class members.

An accurate determination of bulk density starts with proper sampling. Acceptable procedures ensure that the collected sample represents the soil's natural condition and that no compaction or crumbling has occurred. Methods are presented here that illustrate using a simplified tin can sampler, a liquid displacement technique, and a sampling device designed for use by soil scientists.

Part I.   Tin can sampling method

This method is the least accurate of those described, but provides a reasonable estimate of soil bulk density in the absence of more elaborate equipment. It works well in all but stony or loose, sandy soils.

Equipment List

- *Shallow tin can with both ends removed*
- *block of wood*
- *mallet*
- *shovel*
- *large knife*
- *sample bag*

Procedure:

1. With the shovel and/or knife, smooth the surface of the area to be sampled. Select a site where roots are not present.

2. Place the tin can on the soil surface. Put the block of wood on top of the can and hammer the block, forcing the can squarely into the soil. Stop hammering before any compaction occurs.

3. Excavate the buried can and trim excess soil from the bottom until soil is flush with the rim of can.

4. Extract the soil from the can into the labeled sample bag.

5. Dry the soil sample to a constant weight using either microwave radiation or 24 hours in a hot-air oven at 105 °C. Record the dry weight of the soil and calculate the volume of the can (*volume* = $\pi r^2 h$). Calculate the bulk density of the sample.

Part II. Liquid displacement using saran-coated clod technique

An irregular chunk of soil may be used to determine bulk density by employing **Archimedes' Principle,** which states that an object placed in a liquid is buoyed up by a force equal to the weight of the displaced liquid. The basis for this method is measuring a soil's mass (weight) in air and again in water to determine its volume. The liquid displaced by a nonwetting chunk of soil equals its volume, and since water has a density of 1.0 g cm^{-3}, the weight of the clod when suspended in water can be converted to clod volume. Coating the clod with liquid saran will make it nonwetting. This method can be quite accurate because sampling prevents any soil compression, and it can be used in stony soils where other extraction techniques are impossible; but, the method does not work well with loose, sandy soils. Several other substances have been used to seal the clod against water including paraffin. If you use paraffin, accurately determine its density and make sure the paraffin is not too hot when you dip the clods.

Equipment List

- *Airtight sample container*
- *shovel*
- *knife*
- *sewing thread*
- *liquid saran (Prepare by dissolving 1 part by weight of saran resin (Dow Saran F-310; Dow Chemical Co., Suite 500/Tower No. 2, 1701 West Golf Road, Rolling Meadows, IL 60008) in 7 parts by weight of methylethyl ketone. Dissolution requires about 1 hour with vigorous stirring.)*

Procedure

1. Remove a chunk of soil from a desired sampling spot, trimming it to a nearly spherical shape about 4 cm diameter. Handle this piece carefully so that it remains intact and retains its original structure.

2. Tie a sewing thread sling around the sample, leaving a looped end of about 20 cm.

3. Place tied clod into sample container on top of some loose soil to protect it from breaking while transporting it back to the lab. The loose soil should be the same material as your sample since it will be used for moisture determination. Keep container airtight to prevent moisture loss.

4. Remove the intact sample from its container (reclose container to prevent drying) and quickly weigh by suspending it in air from a triple-beam balance equipped with a hook on the balance arm.

5. Holding the thread loop, dip sample into liquid saran for about one second. Dry the coated clod for 5 to 10 minutes in a hood to evacuate the evaporating solvent. Repeat dipping and drying one or more times as needed to waterproof the clod.

6. Weigh the saran-coated sample again by suspending it from the triple-beam balance hook. The difference between the coated and uncoated weights represents the weight of the liquid saran coating.

7. Position a beaker of water such that a third weighing can now be made while the sample is completely suspended in water. If bubbles appear on the saran when the sample is weighed in water, water is penetrating the clod and the sample should be discarded. The difference between weights of the coated sample in air and water will be used to calculate the volume of the coating and soil. The volume of the soil is obtained by subtracting the volume of the coating. (The density of this liquid saran mixture is approximately 1.3 g cm^{-3}, but can be determined more accurately if desired).

8. Finally, to determine moisture content of the clod, weigh a quantity of the loose soil from the sample container and dry to a constant weight using microwave radiation or for 24 hrs in a hot-air oven at 105 °C.

Part III. Metal ring sampler technique

Soil scientists use sampling equipment especially constructed to remove undisturbed soil cores. There are several models available, but all work on the principle of forcing a metallic ring of known volume into the soil. The critical feature of these instruments is the design of the cutting head, which prevents compaction during sampling (see Fig. 4-2). Your instructor will demonstrate the correct sampling technique for the particular device available. Bulk density is then determined by the method presented in Part I.

Part IV. Approximate particle density

Particle density (as well as bulk density) of a sand can be easily approximated in the laboratory by the following procedure.

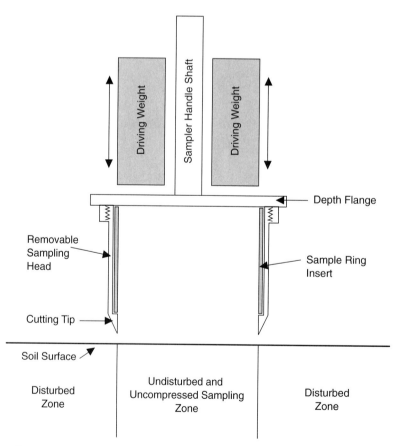

Figure 4-2. Samplers for removing undisturbed soil cores use a cutting tip that wedges soil away from the collection ring and a depth flange to prevent compaction. Unscrewing the cutting head allows extraction of the collection ring containing an undisturbed soil sample.

Equipment List

- *Graduated cylinder (100 mL)*
- *coarse sand*
- *water*

Procedure

1. Weigh 50 g of dry sand and use a funnel to quantitatively transfer to a 100-mL graduated cylinder.

2. Carefully tap the cylinder 4 times to settle the sand. Read the volume and record on your data sheet. Calculate the bulk density.

3. Transfer the sand to a container and save.

4. Add approximately 60 mL of water to the 100 mL graduated cylinder. Record the exact water volume (assume the density of water is 1 g cm^{-3}).

5. Transfer the 50 g of sand from Step 3 back into the cylinder. Stir to remove the trapped air.

6. Read and record the volume. Note the difference between this volume and that in Step 4; this difference is the *volume of the sand particles*.

7. Calculate the particle density by dividing the weight of the sand (50 g) by the volume of the sand particles.

8. Calculate the pore space (% porosity).

Bulk Density and Soil Porosity

Name _____ Section _____

Data

Record data from other class members to compare bulk density and porosity of soils collected from various sites

	Site Description	Texture, Determined from Exercise 2 or 3	Bulk Density, g cm^{-3}	Porosity, %
1				
2				
3				
4				
5				
6				
7				
8				

Parts I. and III. Tin can method or metal ring technique

1. Volume of sampling ring ($V = \pi r^2 h$), cm^3 _____

2. Weight of drying container, g   _____

3. Weight of oven-dry soil + container, g   _____

4. Weight of oven-dry soil, g   _____

5. Bulk density, g cm^{-3} _____

6. Porosity, %   _____

Bulk Density and Soil Porosity

Name _____ Section _____

Data

Part II. Liquid displacement using saran-coated clod technique

1. Soil weight; uncoated clod in air, g _____

2. Soil weight; coated clod in air, g _____

3. Weight of coating, g (line 2 − line 1) _____

4. Density of coating, g cm^{-3} <u>1.3</u>

5. Volume of coating, cm^3 (line 3 ÷ line 4) _____

6. Soil weight; coated clod in water, g _____

7. Volume of soil + coating, cm^3 (line 2 − line 6) _____

8. Volume of soil, cm^3 (line 7 − line 5) _____

9. Loose soil weight; moist, g _____

10. Loose soil weight; dried, g _____

11. Percent water, %$_w$, (line 9 − line 10) ÷ line 10 _____

12. Dry weight of coated soil sample, g (line 1 − [line 1 × line 11]) _____

13. Bulk density, g cm^{-3} (line 12 ÷ line 8) _____

14. Porosity, % _____

Bulk Density and Soil Porosity

Name _____ Section _____

Data

Part IV.   *Particle density technique*

1. Weight dry sand, g   _____

2. Volume of dry sand, cm^3 _____

3. Bulk density of sand, $g\ cm^{-3}$ _____

4. Volume of water, cm^3 _____

5. Volume of water and sand solids, cm^3 _____

6. Volume of sand, cm^3 _____

7. Particle density of sand, $g\ cm^{-3}$ _____

8. Porosity, %   _____

Bulk Density and Soil Porosity

Name _____ Section _____

Questions

1. From your results:

    a.   How does bulk density vary with the amount of clay in the soil?

    b.   How does the porosity vary with the amount of clay in the soil?

2. How does bulk density appear to be related to soil texture? Did the porosity values you found agree with expected relationships to soil texture? If not, point out discrepancies and explain possible reasons for these results.

3. Why will bulk density vary between soils, but it is valid to use a constant value for particle density?

4. How does a change in bulk density affect soil porosity?

5. How are air exchange and water movement in soils related to the amount of pore space? Include the role of pore size distribution in your answer.

Bulk Density and Soil Porosity

Questions

6. How would total pore space and pore size distribution change with depth in a soil that has a sandy loam surface texture and a silty clay loam subsoil?

7. How permanent are bulk density and porosity values? What might cause them to change?

8. In what ways can the bulk density and porosity of a region's soils be instrumental in affecting environmental quality of that locale?

Exercise 5: Soil Water Content

> *Soil acts like a reservoir for water in that it's capable of collecting, storing, and releasing water. Many soil properties regulate water collection, storage, and release, and in turn, the water content of the soil influences the expression of other soil properties. Being able to describe and evaluate soil water content provides tools for conserving water, improving efficiency of water use by plants, and applying appropriate environmental practices.*
>
> *Exercise Goal: You will learn the terms, calculations, and procedures needed to express soil water content and examine the effects of texture and stratification on soil water content.*

Soil water content influences most soil processes and plant growth. Fluctuating continually in response to wetting and drying forces, soil water content is one of the most changeable soil properties. Weather, soil use, management activities, plant growth, and landscape features all influence water balance in soils. Understanding the principles of describing and measuring soil water content is essential to water conservation, including its efficient use in recreation and in food, feed, and fiber production. In addition, because water added to soil contains an array of ingredients from its previous use and exposure, being able to describe its properties and behavior in soil is imperative to sound environmental management.

Soils collect, store, and release water. Collection begins when water enters the soil through surface pores, a process called **infiltration.** Water then permeates throughout the porous soil in patterns dictated by the relative attractive forces of its surroundings. Strong adhesive forces from soil particles and cohesive forces between water molecules retain water in the soil body against the pull of gravity in all but the largest of soil pores. Water adhering to soil particles is under tension because its free energy (matric tension) has been reduced. Tension is strongest (matric potential is lowest) at the soil-water interface and decreases toward the outer boundary of the water layer (Fig. 5-1). Water storage is possible when the forces of retention within the soil exceed removal forces. Water release occurs when plant uptake, drying, or gravitational forces overcome water retention.

Soil Water Energy Benchmarks

Soil water status can be reported on either an energy or content basis. The four benchmarks described in Figure 5-1—field capacity, wilting point, hygroscopic coefficient (air dry), and oven dry—all define the energy status of soil water as matric potential (KPa) or tension (bar). A **saturated** soil would theoretically have water filling all of the soil's pore volume. When a soil is retaining all the water it can against the pull of gravity, it is at **field capacity.** In medium-textured soils, about 50% of pore volume is filled with water at field capacity. When the water film on soil particles is so thin that the energy exerted by plants fails to extract water in sufficient quantities to prevent wilting, the soil is at the **wilting point.** When evaporative forces have removed water down to a layer held with a tension of 3.1 MPa (31 bar), the soil is at the **hygroscopic coefficient** or **air dry** condition. Heat energy equivalent to 1000 MPa (10,000 bar) is required to remove the most tightly bound water molecules and create an **oven dry** soil that will not lose additional water with further heating.

These benchmarks provide a conceptual aid to soil water relations, but routine soil water management usually relies more on approaches that determine soil water content. The relationship between water energy

Soil Water Content

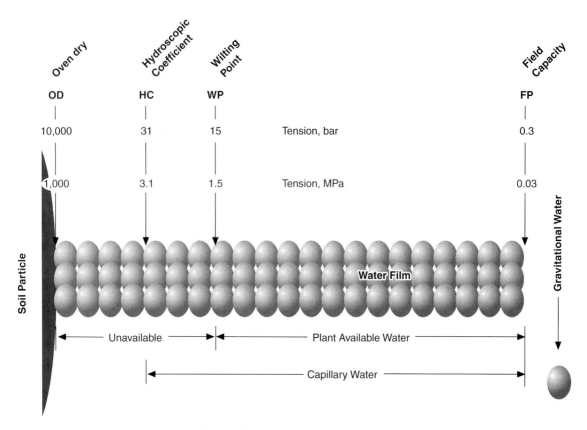

Figure 5-1. The film of water surrounding soil particles is retained by forces that decrease with distance from the particle's surface. Adhesion binds the water molecules nearest the adsorptive surface with a force of 1000 MPa (10,000 bar) tension. When adhesive and cohesive forces decrease below 0.03 MPa (0.3 bar), gravity drains water from soil.

and water content in a soil can be illustrated with a **water retention curve** (Fig. 5-2). Because of differences in composition, each soil will exhibit a unique relationship between soil water energy and soil water content.

Describing Soil Water Content

Water content of soil can be expressed as **percent by weight, percent by volume,** or **equivalent surface depth.** In addition, water depletion can be determined and expressed as either **percent available water depleted** or **field capacity deficit.** Each of these is a different visualization of how much water is either present or has been depleted. However, none of them can describe the energy status of soil water unless a water retention curve is available for that soil.

Water content of a soil can be determined by a **gravimetric analysis** before and after drying and expressed as a percent of the oven dry weight of the soil. This method first requires that a representative sample be obtained and weighed while in a field moist condition. Then the sample is dried at 105°C until its weight remains constant, a condition called the oven dry weight. The weight loss during drying represents soil water, and its percentage composition is expressed as:

$$\text{Percent soil water by weight } (\%_w) = \frac{\text{Wet weight} - \text{Oven dry weight}}{\text{Oven dry weight}} \times 100 \qquad Eq.\ 5\text{--}1$$

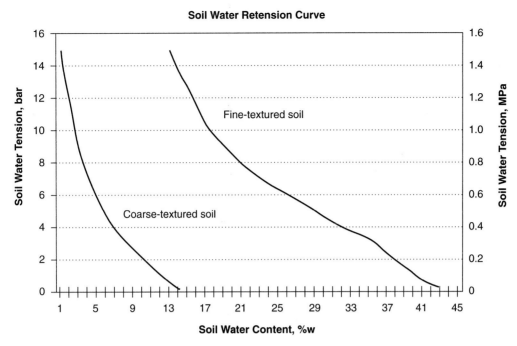

Figure 5-2. A soil water retention curve illustrates the relationship between soil water content and soil water tension. Each soil has a unique curve based largely on clay and organic matter content.

Percent water by weight can be converted to water content expressed as percent water by volume if the soil's bulk density is known. This conversion can be stated as:

$$\text{Percent soil water by volume } (\%_v) = (\%_w)(\text{Bulk density}) \qquad \text{Eq. 5–2}$$

Equation 5-2 is a simplification of the expressions shown here that illustrate the unit cancellations necessary to express percent water by volume as cm^3 water per cm^3 soil.

$$\text{Percent soil water by volume } (\%_v) = \frac{(\%_w)(\text{Bulk density})}{(\text{Density of water})} \times 100$$

$$\%_v = \frac{\left(\dfrac{\text{Wet weight} - \text{Oven dry weight}}{\text{Oven dry weight}}\right)\left(\dfrac{\text{Soil weight}}{\text{Soil volume}}\right)}{\dfrac{\text{Water weight}}{\text{Water volume}}} = \frac{\left(\dfrac{\text{Water, g}}{\text{Soil, g}}\right)\left(\dfrac{\text{Soil, g}}{\text{Soil, cm}^3}\right)}{\dfrac{\text{Water, g}}{\text{Water, cm}^3}} = \frac{\text{Water, cm}^3}{\text{Soil, cm}^3}$$

The percent water by volume expresses the volume of water within a stated volume of soil but, as with percent water by weight, yields no information about the energy status of the soil water unless an applicable water retention curve is available.

Volumetric water values can be converted to represent water content as an **equivalent surface depth.** This expression describes soil water content by how deeply it would cover the soil if removed and set on top of the sample (i.e., the centimeters of rain or irrigation needed to equal this amount of water). Expressing water content as an equivalent depth aids in the visualization of available water deficits and allows comparison to rainfall amounts given in weather reports.

Equivalent surface depth (cm water per sample zone) = $(\%_v)$ *(sample thickness, cm)*   Eq. 5–3

For water management considerations, interest centers primarily on the available water fraction. Irrigation decisions are frequently based on knowing the portion of the available water that has been depleted. A rule of thumb suggests that irrigation should begin when 50% of the available water has been depleted. This value can be calculated as follows:

$$\text{Available water depleted, \%} = \frac{\text{Field capacity}(\%_w) - \text{Current water}(\%_w)}{\text{Field capacity}(\%_w) - \text{Wilting point}(\%_w)} \times 100 \qquad Eq.\ 5\text{–}4$$

The available water depleted value suggests when water replenishment should begin; it does not provide information about how much water should be added. That value can be determined by calculating the **field capacity deficit,** or irrigation requirement, which is the equivalent surface depth of water needed to raise the current soil water content to field capacity. It is calculated as follows:

Field capacity deficit, cm = *(Field capacity, %v − Current water content, %v) (sample thickness, cm)*
                                                                                                                                                         Eq. 5–5

Effect of Soil Texture on Water Content

Soil texture affects water content through its influence on binding sites and storage volume. Water binds to the surface of soil particles, so those soils with the largest specific surface area (surface area per unit mass) have the greatest potential for storing water. Specific surface area is directly proportional to a soil's clay and/or organic matter content (see Exercise 2, Soil Texture). The water-holding ability of sandy soils can be increased by adding residues that raise organic matter.

Soil texture also impacts water content through its influence on storage volume as characterized by aggregation and soil porosity. Water enters soil through surface pores, moves through internal pores, and once bound to soil particles, resides in soil pores. The smallest pores within a soil exert the greatest tension on water and so would fill first and empty last. At field capacity, all but the largest soil pores are filled with water. Loss of aggregation and compaction can diminish a soil's pore volume and, as a consequence, reduce infiltration, volumetric water storage capacity, and water movement and alter patterns of water distribution.

Water added to a soil in excess of its infiltration capacity will pond or leave the site as runoff. If the storage capacity of a soil is exceeded, solutes will percolate into the groundwater. Either condition can spread pollutants.

Effect of Soil Layering on Water Content

The water storage capacity of a soil can be altered by layering within the profile. When texture changes abruptly, so do the tension and porosity properties that affect water distribution. A comparison of these properties is given in Table 5-1. Some layering occurs naturally, whereas other layering is added to a site intentionally or unintentionally and affects soil water content.

Table 5-1. Distribution of water in soil is highly dependent on storage sites (surface area), conduits (pores), and retentive forces (tension) generated by the mix of sand, silt, and clay particles.

Textural layer	Predominant particle and pore size	Specific surface area	Tension generated	Rate of wetting from adjacent layer	Rate of water transmission to adjacent layer
Coarse	Large	Low	Low	Slow	Fast
Fine	Small	High	High	Fast	Slow

Soil Water Content

Laboratory Activity

Part I. Water content and distribution in a soil profile

Several physical and hydrological properties can be determined from collecting a soil sample's volume, moist weight, and dry weight. This exercise illustrates data collection and calculation of percent water by weight, percent water by volume, bulk density, total pore space, and the equivalent depth of water contained in a sample.

Determining Water Content in a Soil Profile.

1. Extract a core sample from a soil using a hand- or power-operated sampling tube. Wrap the core in plastic to prevent water loss between sampling and weighing.

2. In the lab, label sufficient covered containers to accommodate 3-cm segments of the profile (0–3 cm, 3–6 cm, 6–9 cm, etc.). Weigh and record the empty weight of each container and lid.

3. Unwrap the core and quickly measure and record its diameter. Quickly section the core into 3 cm segments; place each segment into the appropriately labeled container and cover to prevent any water loss. Weigh and record the weight of container and wet soil.

4. Crumble the soil for faster drying and place the open container into either a microwave or hot-air oven (105°C). Remove samples at suggested intervals, close the container, and weigh. Continue drying and weighing until a constant weight is attained. At that point the soil sample is losing no more water and is considered to be oven dry. Record the weight of the container and dry soil.

- Microwave soil drying: *Microwaves are a form of high-frequency radio energy waves. The waves are either reflected, pass through, or absorbed by a substance. Metals reflect microwave radiation, but it passes through paper, glass, pottery, wood, and plastic. Water absorbs this energy, causing an increase in molecular vibration; in other words, it heats. When sufficient energy has been absorbed, liquid water evaporates as steam. Either the microwave energy level or time of exposure to a constant energy level can be varied to supply sufficient energy to evaporate all the water from a soil sample. Drying soils to a constant weight requires several minutes in a microwave oven compared to several hours in a hot-air oven. For estimates of drying times for soils, see: Miller, R. J., R. B. Smith, and J. W. Biggar. 1974. Soil water content: Microwave oven method. Soil Sci. Soc. Amer. Proc. 38:535–537. Microwave drying can affect results from some soil analyses; see: Thien, S. J., D. A. Whitney, and D. A. Karlen. 1978. Effect of microwave radiation drying on soil chemical and mineralogical analysis. Commun. in Soil Sci. and Plant Anal. 9(3):231–241.*

- Hot-air soil drying: *Exposing soils to heat energy above the boiling point of water will eventually remove all soil water. Again, water absorbs this energy, causing molecular vibration to increase until the liquid water is driven off the soil particles as a gas. Dry to a constant weight.*

5. Calculate the information requested on the Data Sheet, Part I.

Part II. Effect of soil texture on water content

A. Maximum water retention

Note: this procedure does not equal saturation because some water is allowed to drain; neither does it equate to field capacity because the tension in this soil column is too short to equal the 0.333 cm column of water needed to reach 0.03 MPa (0.3 bar).

1. Prepare soils with different textures by air drying and sieving through a 2 mm screen. These same soils will be used for maximum water retention, hygroscopic coefficient, and wilting point determination.

2. Prepare wetting columns from plastic pipe (approximately 15 cm long with a 5 cm diameter). Secure fine mesh screening over one end of the column with a rubber band. Suspend each tube over a drainage pan. Label each column for the soil textures being used.

3. Put an equal mass of each soil into each column.

4. Add water to each soil column until water starts dripping from the bottom of the column.

5. When water stops dripping from a column, the soil is considered to have maximum water retention. Remove the soil from each column onto a nonadsorbent surface.

6. For each texture, quickly mix the extracted soil and put approximately half into each of two previously weighed beakers, one labeled for oven drying and one labeled for air drying. Quickly reweigh the beaker slated for oven drying before appreciable water loss can occur and record the data. Weigh and set aside the other beaker for determination of air dry water content. Use Data Table II-A.

7. Place the appropriate beaker and wet soil in an oven and dry to a constant weight (see Part I for drying details).

8. Determine the water content of each soil at maximum water retention.

B. Hygroscopic coefficient (air dry)

1. The samples labeled for air drying from Part II-A, Step 6 of the field capacity determination can be placed on the counter at room temperature to air dry. (Avoid areas that have large temperature or humidity fluctuations.)

2. The samples should air dry in four to seven days. The soil should be stirred several times during this interval to facilitate drying.

3. Weigh the beaker and soil when the air dry condition is reached. Then oven dry each sample to a constant weight (see Part I for drying details). Use Data Table II-B.

4. Determine the water content of each soil to approximate the hygroscopic coefficient.

C. Wilting point

Note: Start this exercise several weeks before the remainder of the exercise is scheduled.

1. Put an equal amount of each of the soils into separate, 6-inch plastic pots. Fill pots by adding small amounts of soil and tapping the pots on the benchtop an equal number of times between each addition to settle the soil. Remove approximately 1 inch of the soil and sow about 20 seeds (wheat, oats, or barley work well) onto the soil and cover the seeds with the soil previously removed.

2. Place the pots in a greenhouse, water, and allow the plants to grow until the root systems thoroughly permeate the soil, then withhold water from the pots. The soil will be at the wilting point when plants wilt and are not revived by misting the leaves with water. (Cover the soil surface when misting to prevent water entry into the soil.)

3. When the soil is determined to be at the wilting point, remove the soil from the pot and shake free as much of the root system as possible. Quickly mix the root-free soil and place a sample into a previously

weighed and labeled beaker. Reweigh the beaker and soil before appreciable water loss can occur and record the data. Use Data Table II-C.

4. Place the beaker and soil in an oven and dry to a constant weight (see Part I for drying details).

5. Determine the water content of each soil at the wilting point.

Part III. Effect of soil layering on water content

Layers within a soil profile can affect the water-holding capacity of the overlying soil. When coarse particles underlie finer particles, the water retention of the overlying layer is increased. The larger pores and coarser particles of the underlying soil cannot exert as much tension on soil water as the overlying soil does. Hence, water reaching the interface of those layers is retained in the overlying layer until it becomes quite saturated (i.e., reaches a low tension). Soil constructed in this manner, for example, golf greens, allows sandy root zones to retain more water than if the underlying gravel layer was omitted.

When fine particles underlie coarser particles, the water retention of the coarser soil is reduced. In this case, the high tension in the fine-textured layer attracts water away from the low tension (coarse) layer.

1. Place a 1- to 2-inch thick layer of fine gravel (2–25 mm diameter) on the screen of a medium to fine mesh sieve pan. Suspend the sieve pan on pedestals in a shallow basin (upside down beakers work well as pedestals). The gravel creates a coarse pore layer with low tension.

Prepare a second unit in a similar fashion, except use a 2-inch thick layer of silt loam soil instead of gravel. The silt loam creates a fine pore layer with comparatively high tension.

2. Label 12 1-inch-thick sponges and record their individual dry weights. Create a "sponge profile" by stacking 6 sponges on the gravel base and 6 sponges on the soil base.

3. Slowly pour water from a graduated cylinder so that it enters the top sponge of the stack setting on the gravel layer. Keep adding water slowly and evenly across the top sponge to thoroughly wet all sponges. Stop adding water to the top sponge when water starts dripping from the gravel layer. Record the amount of water added.

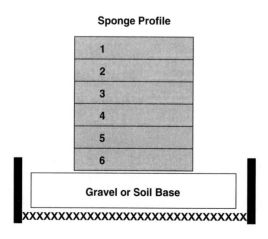

4. Allow the sponge profile to continue dripping into the gravel layer. When it stops dripping, the "profile" is near saturation.

5. Use the same amount of water as required in Step 3 and slowly wet the sponges atop the soil base. Allow time for the water to equilibrate throughout the sponges and reach saturation.

6. When both profiles have reached saturation, dismantle the sponge profile and place each sponge into a previously weighed beaker and reweigh. Handle sponges gently so no water is lost during transfer.

7. Dry each sponge to a constant weight (see Part I for a drying method).

8. Compute the water content of each sponge and answer the associated questions.

Part IV. Water calculations

As population pressures put increasing demands on global water supplies, the necessity to quantify water use becomes more important to efficient management. Drawings like Figure 5-3, depicting the water content at principal benchmarks, can help visualize the different classes of soil water.

1. Use the equations in this exercise and in Figure 5-3 to fill in the data sheet for Part IV.

2. Using the data for the top layer of soil in the data sheet for Part IV, prepare a drawing like that in Figure 5-3.

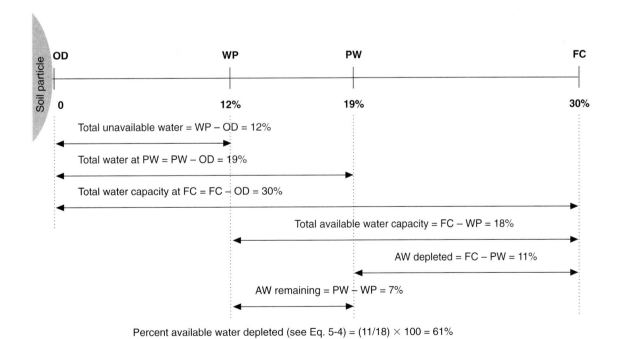

Figure 5-3. A visual depiction of soil water content at benchmark values and their use in determining quantifiable water information. (OD = oven dry, WP = wilting point, PW = present wetness, FC = field capacity, AW = available water).

Soil Water Content

Name _____ Section _____

Data

Part I. Water content of a soil profile

	Soil Profile Segment, cm							
	0–3	3–6	6–9	9–12	12–15	15–18	18–21	21–24
Diameter of core, cm								
Volume of core, cm³ (V = $\pi r^2 h$)								
Weight of container and lid, g								
Weight of wet soil, container, and lid, g								
Weight of dry soil, container, and lid, g								
Weight of water in soil, g								
Oven dry weight of soil, g								
Percent water by weight, $\%_w$								
Bulk density, g cm^{-3}								
Percent water by volume, %								
Equivalent surface depth, cm								

Name _____ Section _____

Data

Part II. *Effect of soil texture on water content*

	A. Maximum water retention					
	1	2	3	4	5	6
Soil texture						
Weight of wet soil and beaker, g						
Weight of beaker, g						
Weight of oven dry soil and beaker, g						
Weight of water in wet soil, g						
Oven dry weight of soil, g						
Percent water by weight at maximum water retention, $\%_w$						

Soil Water Content

Name _____ Section _____

Data

Part II. Effect of soil texture on water content

	B. Hygroscopic coefficient (air dry)					
	1	2	3	4	5	6
Soil texture						
Weight of air dry soil and beaker, g						
Weight of beaker, g						
Weight of oven dry soil and beaker, g						
Weight of water in soil at air dry condition, g						
Oven dry weight of soil, g						
Percent water at air dry, $\%_w$						

Name _____ Section _____

Data

Part II. Effect of soil texture on water content

	C. Wilting point					
	1	2	3	4	5	6
Soil texture						
Weight of wilting point soil beaker, g						
Weight of beaker, g						
Weight of oven dry soil and beaker, g						
Weight of water in soil at wilting point, g						
Oven dry weight of soil, g						
Percent water when soil is at wilting point, $\%_w$						

Soil Water Content

Name _____ Section _____

Data

Part III. Effect of soil layering on water content

Coarse Underlying Layer						
Sponge No.	1	2	3	4	5	6
Weight of wet sponge and beaker, g						
Weight of beaker, g						
Weight of oven dry sponge and beaker, g						
Oven dry weight of sponge, g						
Weight of water in sponge, g						
Percent water, $\%_w$						
*Percent water of profile, $\%_w$						

Fine Underlying Layer						
Sponge No.	1	2	3	4	5	6
Weight of wet sponge and beaker, g						
Weight of beaker, g						
Weight of oven dry sponge and beaker, g						
Oven dry weight of sponge, g						
Weight of water in sponge, g						
Percent water, $\%_w$						
*Percent water of profile, $\%_w$						

*$\dfrac{\textit{Sum of weights of water in all sponges, g}}{\textit{Sum of dry weights of all sponges, g}} \times 100$

Name _____ Section _____

Part IV. Water calculations

Use the information in Columns 1–4 to calculate values in Columns 5–11

Depth cm	Dry weight, g	Wet weight, g	Sample volume, cm³	Wilting point	Field capacity	Soil Profile	Present water content	Equivalent surface depth, cm	Accumulated surface depth, cm	Available water depleted, %	Field capacity deficit, cm
1	2	3	4	5	6		7	8	9	10	11
0–15	424	509	320	12%$_w$ 16%$_v$	40%$_w$ 53%$_v$		%$_w$ %$_v$				
15–30	510	627	364	14%$_w$ 20%$_v$	42%$_w$ 59%$_v$		%$_w$ %$_v$				
30–45	500	685	345	16%$_w$ %$_v$	48%$_w$ %$_v$		%$_w$ %$_v$				
45–60	319	431	220	19%$_w$ %$_v$	30%$_w$ %$_v$		%$_w$ %$_v$				
60–75	468	571	347	10%$_w$ %$_v$	30%$_w$ %$_v$		%$_w$ %$_v$				
75–90	414	455	323	9%$_w$ %$_v$	25%$_w$ %$_v$		%$_w$ %$_v$				

Soil Water Content

Name _____ Section _____

Questions

1. How does a saturated soil differ from a soil wet to field capacity?

2. How does an air dry soil differ from an oven dry soil?

3. How can knowledge of the energy status of soil water be converted to soil water content?

4. A 400 cm³ sample from the top 25 cm of a soil weighed 686 g when collected and 576 g when oven dry. This soil contains 34%$_w$ water at field capacity. What is the bulk density? What is the percent water? What is the field capacity deficit of this soil zone?

5. What does it mean to dry a soil to a constant weight?

6. Explain the relationship between soil texture and water content at field capacity.

Soil Water Content

Name _____ Section _____

Questions

7. Describe the relationship between soil texture and plant available water.

8. How could the plant available water capacity of a soil be increased?

9. How did the presence of a coarse or fine pore layer affect the water content of the overlying sponge profile?

10. Was the water content of the six sponges equally distributed at field capacity? How did the water content differ in the sponges? Why did this difference exist?

11. Describe a condition where the principles illustrated in this exercise could be applied to (a) increase the water-holding capacity of a soil, and (b) reduce the potential for water distribution of a soluble environmental hazard.

Exercise 6: Water Movement in Soil

> *The distribution of water in soil has important implications for plant growth and environmental considerations. Water added to a site will either runoff, pond on the surface, or infiltrate into and redistribute within the soil body. In either case, where the water resides and the types of constituents it carries affect soil use and resource conservation.*
>
> *Exercise Goal: In this exercise you will examine the types of water movement within a soil profile that determine the wetting patterns typically hidden from view. You will see a video and perform simple lab activities that illustrate important principles of water movement in soil.*

The distribution of water in soil is generally not uniform, nor is its behavior easy to understand. Besides being essential for plant growth, water transports a vast array of dissolved and suspended constituents that further impact plant growth, soil use, and environmental considerations. Whether water reaches a site as rainfall, irrigation, surface flow, or ground water, its redistribution within the soil body determines its ultimate impact. This exercise studies those soil properties and principles that influence water movement in soil.

Forces Affecting Soil Water Movement

Water moves in soil in response to three forces: gravity, evaporation, and tension. **Gravity** attracts water downward through the soil body, but only has influence if the other forces are small or negligible. **Evaporation** occurs when meteorological elements produce sufficient energy to convert liquid water into a gas that moves from the soil body to the atmosphere. **Tension** describes the attraction of water to other entities. The attraction of a water film onto solid soil particles is called **matric tension.** Matric tension is strongest for water molecules nearest soil particles and weakens progressively toward the outer regions of the film. Water molecules are attracted to dissolved salts by a force called **osmotic tension.**

In all situations, water obeys simple rules of physics by moving from a zone where attractive forces are small toward a zone where those forces are larger (Fig. 6-1). Pressure (positive values) and tension (negative values) forces can be measured with atmosphere (atm), bar, or Pascal (Pa) units. **Unsaturated** conditions occur when the soil water is under tension. If soils become so completely **saturated** that all tension forces have been satisfied, then positive water pressures can exist. When the pressure scale is applied to soil conditions, water movement will only be from high pressure zones to low pressure zones (i.e., from right to left in Fig. 6-1).

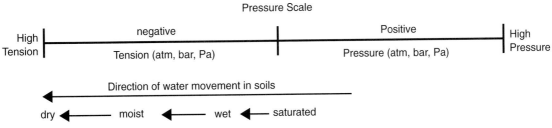

Figure 6-1. Components of the pressure scale related to direction of water movement in soils.

Unsaturated Water Flow

Unsaturated flow occurs when the water film held onto soil particles by tension streams to another region exerting a higher tension. This particle-to-particle film thickness adjustment can occur in any direction as long as a tension gradient exists.

To understand unsaturated flow it is necessary to understand **capillarity,** or water flow in small openings while under tension. Water enters capillaries because of **adhesion,** an attraction between water molecules and molecules in the capillary wall. As water creeps along the capillary walls, other water molecules are pulled along by **cohesion,** the attraction of water molecules for each other. The curved water-air interface formed in capillaries is known as the **meniscus** (Fig. 6-2). Water forming the meniscus

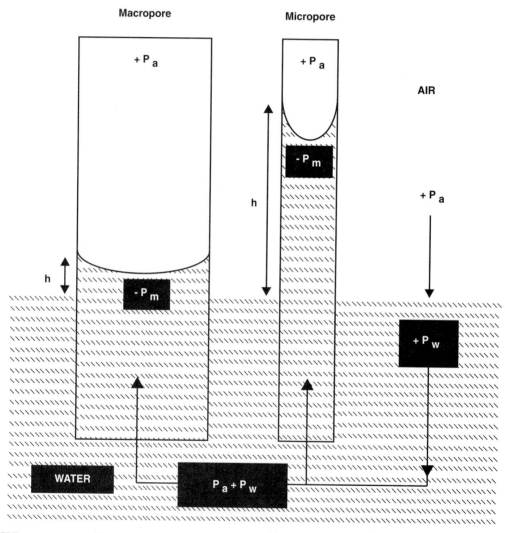

Figure 6-2. When open capillaries are placed in water, there are positive pressures outside the tube that exceed the negative pressure (tension) from adhesion and cohesion inside the capillary. Responding to this imbalance of forces, water moves into the capillary. Small capillaries produce a meniscus with greater tension (i.e. more curvature) than large capillaries, which in turn produces a bigger gradient between forces in and outside the capillary. The larger this gradient in forces, the higher will be the capillary rise.

P_a = atmospheric pressure P_m = meniscus tension
P_w = water pressure h = capillary rise

is under tension and the greater the curvature of the meniscus, the greater will be the tension. The positive pressures shown in Figure 6-2 push water toward the area of negative pressure in the meniscus, causing water to rise in the capillary. Water continues to rise until the difference between the positive and negative pressure is balanced by the weight of water suspended in the column. The inverse relationship between the height water rises in a capillary and the size of the capillary can be expressed by this formula:

$$h = \frac{0.3}{d}$$

where h is the height (cm) of rise of the water column and d is the effective diameter (cm) of the capillary at the air-water interface. The smaller the radius of a capillary, the greater the tension it exerts. For this reason water held in large pores evaporates or is used by plants before water held in small pores. As the soil dries, water is found only in progressively smaller pores and held with progressively higher tension. When water is bound by about 15 bar tension, a permanent wilting point is reached because plants obtain water too slowly to remain turgid.

In soil, capillaries are not of uniform diameter so the exact relationship described in this formula may not always apply, but the relative relationship between water movement and pore size (as related to soil texture) still holds true (Fig. 6-3). It should be noted that since capillarity involves water under tension, the water movement illustrated in Figures 6-2 and 6-3 would still occur if the direction of movement was

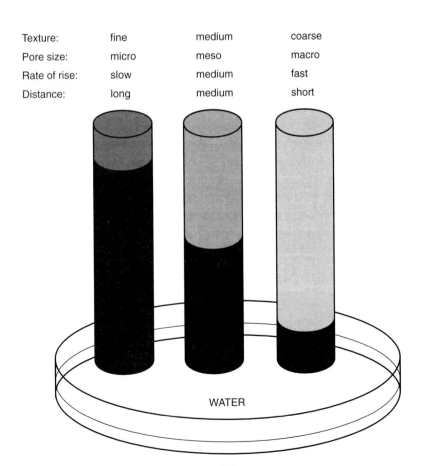

Figure 6-3. Capillary movement in soil columns with different textures.

sideways, or downward. Also, the rate of water movement by unsaturated flow is inversely proportional to the distance it moves from the water source. This means water moves quickly into a macropore but does not rise very high, and conversely, water moves slowly into a micropore but will rise much higher than in the macropore.

Saturated Water Flow

When the soil water content increases past the point that all tension forces are satisfied, conditions of **saturated flow** exist. Water movement no longer responds to forces of attraction from soil particles, but freely obeys the force of gravity. The direction of flow is downward and the rate of flow is proportional to the pore size and **hydrostatic head** (depth of the saturated zone plus any surface water).

Large pores open to a source of surface water facilitate rapid water entry and distribution by saturated flow. However, if these open channels become plugged and cut off from the source of free water, saturated flow ceases until the surrounding soil becomes saturated.

Water Movement from Furrows

Water entering a soil from an irrigation furrow moves radially away from the entry zone (Fig. 6-4). Horizontal and upward movement is nearly as great as the downward movement, a distribution pattern indicating conditions of unsaturated flow. The wetting front expands into the dry soil because the tension

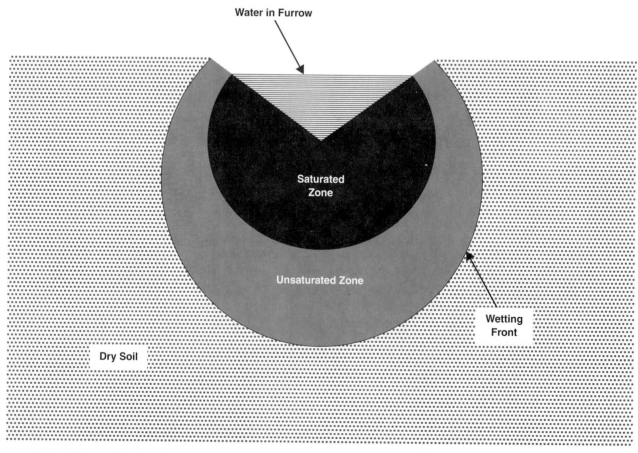

Figure 6-4. Water distribution away from a furrow creates a zone of saturated flow near the water source led by a zone of unsaturated flow out to the wetting front. The saturated zone will soon begin to elongate downward because gravity is the driving force in its movement.

there is higher than in the moist zone. Soluble salts and fertilizers will be carried along with the path of the advancing wetting front. This means soluble salts located above the water level in the furrow will be transported toward the soil surface, away from plant roots, and accumulate there as the water evaporates.

Eventually, a zone below the water source wets sufficiently to satisfy all tension demands and takes on the characteristics of saturated flow. Water in the saturated zone will be moved downward by the force of gravity.

Water Movement in Layered Soil

The principles just illustrated for water movement by saturated and unsaturated flow can be combined with water distribution away from a furrow to illustrate water movement in layered soils. Figure 6-5 illustrates how a coarse-textured layer in a fine-textured soil and a fine-textured layer in a coarse-textured soil influence what would otherwise be a more uniform distribution of water. Layered soils occur in golf

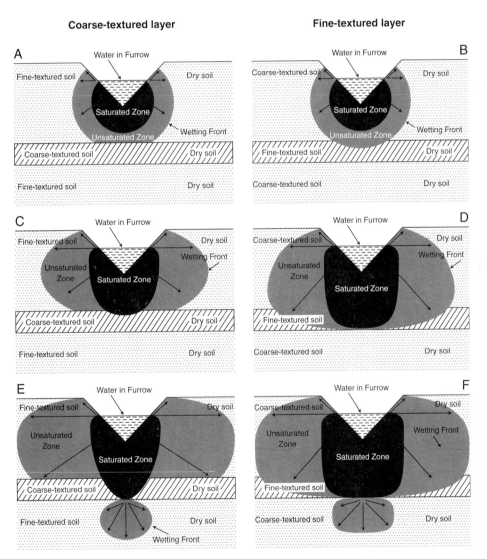

Figure 6-5. Water movement is altered by textural layers within the soil. These drawings illustrate sequential conditions in soil with a coarse- or fine-textured layer. The coarse-textured layer in A, C, and E wets slowly but has a fast rate of water release to the underlying layer. The fine-textured layer in B, D, and F wets rapidly but has a slow rate of water release to the underlying layer.

greens, athletic fields, sodded lawns, landfill sites, and under natural conditions where several deposition factors have influenced soil formation.

Study the sequence of illustrations in Figure 6-5 to find application of the following principles:

- *Water movement in the unsaturated zone always is from an area of low tension to an area of high tension.*
- *Water movement in the saturated zone is always downward, under the influence of gravity.*
- *A coarse-textured layer in a fine-textured soil has less tension than the fine-textured soil and so will not wet until the overlaying soil becomes nearly saturated.*
- *A fine-textured layer in a coarse-textured soil has more tension than the coarse-textured soil and wets as soon as the wetting front reaches it.*
- *A coarse-textured layer (low tension) in a fine-textured soil releases water to the underlying soil (high tension) as soon as it wets.*
- *A fine-textured layer (high tension) in a coarse-textured soil will not release water to the underlying soil (low tension) until it becomes nearly saturated.*

Water Movement in Buried and Open Channels

Tillage, aeration, and other soil management steps often create channels that are filled with residue or sand. These channels aid water movement into the soil only when they are open to a source of free water (Fig. 6-6). Water moves into these channels by saturated flow and from the channels into the soil body by unsaturated flow. When the channels become sealed by fine soil, they impair uniform distribution by unsaturated flow and will not wet until the soil around them becomes saturated. Thus, buried channels wet only after the surrounding soil has become nearly saturated. Only when the filled channels remain open to the soil surface will residue or sand additions increase water infiltration.

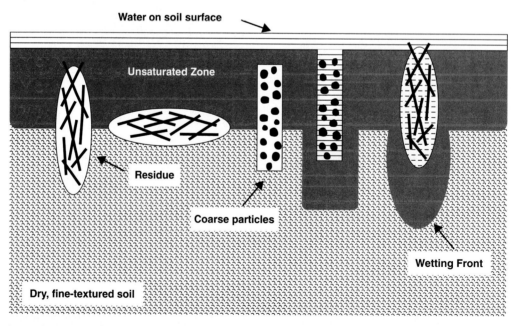

Figure 6-6. Aeration cores filled with sand and partial incorporation of residue facilitate water infiltration by furnishing large surface pores for infiltration. The same material covered by fine-textured soil will not wet until the surrounding soil is nearly saturated.

Effect of Soil Texture on Water Movement

Soil texture modifies water movement through its influence on specific surface area and porosity. Fine-textured soils have higher specific surface area than do coarse-textured soils, which in turn increases water-holding capacity and reduces the rate of water movement. Thus, a clay loam, for example, holds much more water than a sandy loam when both are wet to field capacity, but a wetting front moves more quickly within the sandy loam after an equal water application (Fig. 6-7).

Soil texture influences porosity primarily through its effect on pore size distribution. Macropores dominate in coarse-textured or highly aggregated soils, while micropores dominate in fine-textured or compacted soils. The rate of water movement is directly related to pore size, while the distance water moves is inversely related to pore size.

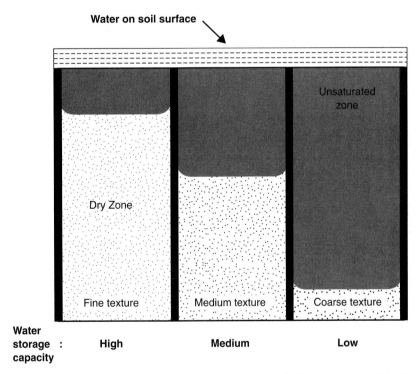

Figure 6-7. Surface water will move into a coarse-textured soil quickly because of the large pores that facilitate saturated flow and the low specific surface area that promotes rapid unsaturated flow.

Laboratory Activity

Part I. Video of water movement

The manner in which several soil properties influence water movement can be seen on video. Study the background information in this chapter and then watch *How Water Moves Through Soil* by Jack Watson (This video was produced in 1994 and is available from the College of Agriculture, University of Arizona, 715 North Park Avenue, Tucson, AZ 85719, Phone: 602-621-7176). The original film, *Water Movement in Soil*, by W. H. Gardner and J. C. Hsieh or the slide set based on this film distributed by Washington State University may also be used for this exercise.

Use the video to observe, define, and understand the following properties and processes and how they affect water movement. Answer the questions that follow in this exercise.

- *Capillarity*
- *Gravity*
- *Unsaturated flow*
- *Saturated flow*
- *Soil texture*
- *Water retention*
- *Soil layering*
- *Adhesion*
- *Cohesion*
- *Pore space and pore size*
- *Direction of solute flow*
- *Open and closed channels*

Part II. Infiltration and capillary rise in soil

Soil texture and layering greatly influence infiltration and capillary rise. This activity will allow quantification of those effects and can be interpreted with the aid of information from Exercise 5, Soil Water Content, and from the other sections of this exercise.

Soil Materials: Prepare soil materials as follows: air dry and sieve through a 1-mm sieve samples of a fine-textured soil, medium-textured soil, coarse-textured soil, and a mixed-texture soil (equal parts of the fine-, medium-, and coarse-textured soil mixed together). Prepare some coarse sand by air-drying and retaining particles larger than 0.5 mm and smaller than 1.0 mm.

Infiltration and Capillary Rise Columns: Prepare twelve 100-mL clear plastic graduated cylinders by drilling eight, 5/16" holes around the bottom for drainage/intake and three columns of 1/16" holes between the 90-mL and 20-mL marks for air escape. Label cylinders as A_i, B_i, C_i, D_i, E_i, F_i, A_c, B_c, C_c, D_c, E_c, and F_c (see diagram) and weigh.

1. Prepare 12 soil columns as illustrated here. The manner in which these columns are packed is critical to the success of this exercise. First, place tape over the bottom drainage holes. Put only 10-mL depths of soil into the cylinder at a time and then firmly tamp the cylinder on the bench three times. Repeat this process until the cylinder is filled to the 100-mL mark, or other mark in the case of layered columns. When cylinders have been filled to the 100-mL mark with soil, remove tape and weigh each cylinder.

2. For those cylinders measuring infiltration (A_i, B_i, C_i, D_i, E_i, and F_i), note the time and begin wetting the soil by gently pouring water onto the soil surface from a beaker. Maintain free water on the soil surface at a depth of 0.5 cm at all times. Record the time the wetting front reaches the 90, 80, 70, 60, 50, 40, 30, 20, 10, and 0 mL marks on the cylinder.

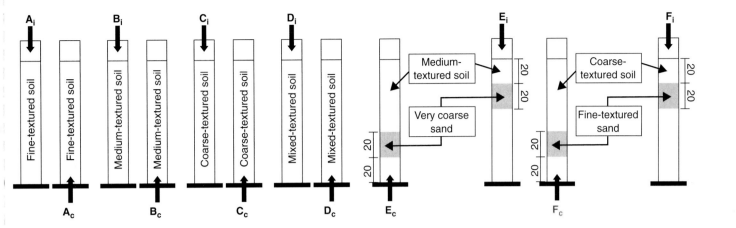

3. For those cylinders measuring capillarity (A_c, B_c, C_c, D_c, E_c, and F_c), note the time and begin wetting the soil by setting the cylinder into a pan of water. Maintain water in the pan at a depth equal to the 5-mL mark on the cylinder. Record the time the wetting front reaches the 10, 20, 30, 40, 50, 60, 70, 80, 90, and 100 mL marks on the cylinder.

4. When the wetting front has traveled its maximum distance, withhold any additional water application (i.e., add no more water to the surface or remove cylinder from pan of water). Let the cylinder drain for 5 minutes. Dry off the outside of the cylinder and weigh. Fill in the data table and answer the accompanying questions.

Part III. *Effect of pore column height on water retention*

Water retention illustrated with screening.

1. Fill a shallow pan with water and lay a square of screen material in the water.

2. Lift the screen out of the water, keeping it horizontal. Note that the "pores" of the screen retain water.

3. Now hold the screen in a vertical position. Note that water drains from the "pores." Interpret these results.

Water retention illustrated with a sponge column.

1. Attach three sponges to a single ring stand using three clamps. Position each sponge so it will not drain onto another sponge.

2. Thoroughly wet each sponge with water and allow to drain until reaching "field capacity", i.e., the point at which the sponge stops dripping.

3. Now adjust the clamps so that the sponges form one column three sponges high. Make sure that the sponges are in contact with each other. Note what happens to the water contained in the sponges. Interpret this behavior.

4. Touch the sponge column at various places. Note which location is the wettest, which is the driest. Interpret these observations.

Part IV. *Capillarity and air displacement*

Air must be displaced from soil pores to accommodate water entry and movement. Aggregates wet by capillary action that begins at the source of the water and progresses away from the source. When a water film initially contacts only a part of the aggregate's outer surface, air in the internal pores can escape ahead of the capillary wetting front. Conversely, if the aggregate is immersed, wetting originates over the entire outer surface of the aggregate and blocks air escape. In that case, air pressure builds in the aggregate as capillarity progresses inward until air eventually (1) bubbles out of the aggregates if they are stable, or (2) "explodes" the aggregates if they are unstable.

1. Collect soil representing tilled, untilled, high clay, low clay, high organic matter content, and low organic matter content sites. Air dry and sieve soils, retaining aggregates between 3 and 10 mm.

2. To illustrate slow wetting, place an aggregate from each soil type onto moist filter paper in a petri dish. Use a low-power binocular scope and observe the aggregates as they wet.

3. To illustrate the effects of rapid immersion, drop an aggregate into a small beaker filled with water. Use a low-power binocular scope and observe an aggregate from each soil type as it wets.

4. Interpret observations to explain how wetting method illustrates differences in aggregate stability.

Name/Group: _____

Data

Part II. Infiltration and capillary rise in soil

	Infiltration						Capillarity					
	A_i	B_i	C_i	D_i	E_i	F_i	A_c	B_c	C_c	D_c	E_c	F_c
Cylinder and dry soil, g												
Cylinder, g												
Dry soil, g												
Cylinder and wet soil, g												
Water in column, g												

mL mark	"depth"	Time (h:mm:ss)	"rise"	Time (h:mm:ss)
90	10		10	
80	20		20	
70	30		30	
60	40		40	
50	50		50	
40	60		60	
30	70		70	
20	80		80	
10	90		90	
0	100		100	

Water Movement in Soil

Name _____ Section _____

Questions

Part I. Water movement video

1. What forces account for water retention in soils?

2. What forces are involved in movement of water in soils?

3. When a fine-textured soil overlays a layer of coarse-textured soil, why doesn't the water immediately flow into the coarse-textured layer when the wetting front reaches it?

4. Under what conditions will water in an overlying fine-textured soil enter a coarse-textured layer?

5. Explain how both a coarse-textured layer and a fine-textured layer can cause a buildup of the water content in the overlying soil.

Water Movement in Soil

Name _____ Section _____

Questions

Part I. Water movement video

6. Compare the saturated flow and unsaturated flow occurring in and near two aeration cores filled with sand; one open to the surface and one covered with a fine soil.

7. Under what conditions will water flow into and through buried pores such as worm holes and root channels that are not open to the surface?

8. When tilling surface residue into a soil, what guidelines are suggested that will maximize water infiltration?

9. Under what conditions will drainage tile remove water from a soil profile?

10. Describe a situation in which application of the principles of water movement will reduce or eliminate an environmental problem.

Water Movement in Soil

Name _____ Section _____

Questions

Part II. Infiltration and capillary rise in soil

1. Using a spreadsheet program or suitable graph paper, prepare two graphs from your data, one for infiltration and one for capillarity. Graph "depth" (or "rise") on the vertical axis and time on the horizontal axis. Plot six lines on each graph representing the six types of soil columns. Include all essential components of a graph: title, axis labels, legends for treatment lines, name, date, etc.

2. Which soil column showed the fastest infiltration rate? Which soil column showed the slowest infiltration rate?

3. What determines water infiltration rate in soil?

4. Explain how the presence of a fine-textured layer and a coarse sand layer affected infiltration rate.

5. Which soil column showed the fastest rate of capillary rise? Which soil column showed the slowest rate of capillary rise?

6. What determines the rate of capillary rise in soil?

Water Movement in Soil

Name _____ Section _____

Questions

7. Explain how the presence of a fine-textured layer and a coarse sand layer affected capillary rise.

8. How are the rate of infiltration and water content of the soil column correlated?

9. How are the rate of capillary rise and water content of the soil column correlated?

Part III. *Effect of pore column height on water retention*

1. Why did the screen's pores stay filled when the screen was held horizontal, but emptied when the screen was held vertical?

2. Why did the sponge column release additional water compared to that released by the three individual sponges?

3. Which part of the sponge column felt the wettest? The driest? Explain the difference.

Name _____ Section _____

Questions

4. Describe a natural situation in which the principles illustrated by the screen and sponge column apply.

Part IV. Capillarity and air displacement

1. How does clay content affect aggregate stability? Why?

2. How does organic matter affect aggregate stability? Why?

3. How does soil management (tilled versus untilled) affect aggregate stability? Why?

4. Describe how water management and aggregate stability affect soil erosion.

Exercise 7: Environmental Influences on Soil Formation

> *Soil properties reflect the accumulation of a multitude of interactions with their environment. Both natural processes and human activities contribute to that environment. A host of natural soil-forming processes, influenced by five soil-forming factors, produce natural paths for soil development. Humans can intervene on a scale sufficient to alter these natural occurrences. The long-term effect of human intervention has yet to be established.*
>
> *Exercise Goal: This exercise will illustrate some specific natural weathering processes and reference how humans might alter the way soils interact with their environment. You will use demonstrations of some specific soil formation processes to illustrate how soils become a product of their environment.*

The accumulative interactions between geologic deposits and their environment determine the current properties of a soil. Natural environmental interactions represent processes that have operated over geologic eras and are still active today in bringing change to soils. Humans, a relatively recent factor important to the soil environment, can intervene in these natural processes of soil formation and alter soil properties. Management that allows a soil to become acid, for example, may accelerate some weathering reactions, or covering large expanses of soil with concrete or asphalt may limit water access and retard weathering processes. Human interactions can improve soil through fertilizing or liming, or degrade soil properties through contamination, depletion, pollution, erosion, and/or compaction.

Over time, environmental interactions transform geologic deposits into soil profiles, with the accumulated change called **differentiation.** The myriad of natural interactions of soils and their environments can be categorized as additions, losses, redistributions, or transformations (Table 7-1). Within each category, many specific types of natural processes have been identified by soil scientists. One example is **weathering,** an interaction with climatic elements that brings physical and/or chemical change to rocks and minerals.

Physical Weathering

Soil formation begins as rock and mineral particles undergo **physical weathering,** a process that reduces a particle's size without changing its chemical composition. Differential thermal expansion during heating and cooling, expansion and contraction during wetting and drying or freezing and thawing, and abrasion by glaciers or wind-borne particles are the dominant agents of physical weathering.

Through particle size reduction, physical weathering increases the surface area available to chemical weathering. Surface area increases by the same factor by which the particle size was reduced. For example, halving the particle size of a given mass doubles the surface area of that mass. The greater the amount of surface area a material exposes to its environment, the faster it can be weathered by chemical processes. Physical weathering will be the most expressive formation process wherever dryness dominates the soil environment.

Table 7-1. These soil processes, along with many others, contribute to differentiation, the conversion of geologic deposits into soil profiles. The five soil-forming factors listed in Table 7-2 function as external controls that determine the intensity of soil-forming processes.

Additions	Deposition: wind or water additions of soil material Littering: accumulation of organic material on the soil
Losses	Leaching: removal of soluble materials Erosion: removal of the surface layer
Redistributions	Eluviation: movement of material out of a zone Illuviation: movement of material into a zone Pedoturbation: biological or physical mixing of soil materials Salinization: accumulation of soluble salts Alkalization: accumulation of sodium
Transformations	Weathering: changes due to exposure to climatic elements Decomposition: breakdown of mineral and organic materials Humification: conversion of organic material into humus Mineralization: release of mineral constituents from organic matter Synthesis: formation of new mineral or organic species

Chemical Weathering

Reactions that change the chemical composition of soil mineral matter are collectively called **chemical weathering.** The five categories of chemical weathering processes are: dissolution, hydration, hydrolysis, carbonation, and oxidation-reduction. Water is a key ingredient in all chemical weathering processes, and the soil biota (roots and microbes) are involved in some, especially carbonation.

- *Dissolution occurs when water causes a cation and anion to dissociate (separate into ions). When forces of attraction within a compound are exceeded by the attractive forces of surrounding dipolar water molecules, the compound dissociates and its ions become part of the solution. Water is a nearly universal solvent, creating, over time, the soil solution.*

$$\underset{Solid}{KCl} \underset{Water}{\rightarrow} \underset{Solution}{K^+ + Cl^-}$$

- *Hydration is the chemical combination of water with another substance. The water molecules do not dissociate, but attach as a shell to the other substance. Hydrated substances exhibit different chemical and physical properties than when unhydrated. The formation of limonite from hematite is a typical hydration reaction in soils and one that causes soil color to change.*

$$\underset{Hematite\ (red)}{2\ Fe_2O_3} + 3\ H_2O \rightarrow \underset{Limonite\ (yellow)}{2\ Fe_2O_3 \cdot 3\ H_2O}$$

- *Hydrolysis begins with the attraction of dipolar water molecules to mineral surfaces and the splitting of the water molecule into H^+ and OH^- ions. The H^+ ions then combine with minerals by replacing soluble* *constituents. In the example reaction, potassium in the microcline is replaced by hydrogen with the formation of an acid form of the mineral and basic potassium hydroxide. The acid alumino-silicate structure is less stable and weathers faster than its microcline precursor does, and the potassium hydroxide is a soluble form of the plant nutrient potassium.*

$$\underset{\text{Microcline}}{KAlSi_3O_8} + HOH \rightarrow \underset{\text{Acid Silicate}}{HAlSi_3O_8} + \underset{\text{Potassium hydroxide in solution}}{K^+ + OH^-}$$

- **Carbonation** *is a set of acidification processes that change soil minerals. First, carbonic acid forms from reaction of water with the carbon dioxide released into the soil solution by root or microbial respiration. Then, basic oxides are converted to carbonates through reaction with carbonic acid. In this illustration of the carbonation reaction, magnesium oxide is converted to magnesium carbonate, a more soluble and plant-available form.*

$$CO_2 + H_2O \leftrightarrow \underset{\text{Carbonic acid}}{H_2CO_3}$$

$$\underset{\text{Magnesium oxide}}{MgO} + H_2CO_3 \rightarrow \underset{\text{Magnesium carbonate}}{MgCO_3} + H_2O$$

- **Oxidation-reduction** *is a set of coincident reactions involving electron transfer between atoms.* **Oxidation** *occurs when an atom loses an electron,* **reduction** *occurs when an atom gains an electron. An electron transfer will change the valence of each atom. Valence changes affect the solubility of the compound containing the oxidized or reduced atom and, in turn, the compound's susceptibility to physical or chemical weathering. The reduction of iron in water-logged (i.e., O_2-deficient) soil and its subsequent oxidation under aerobic conditions typifies this process. Ferrous iron is very soluble and mobile in the soil profile, but exposure to O_2 readily oxidizes it to ferric iron. Ferric oxide is the tan, brown, or red concretions and particle coatings seen in soil.*

$$\underset{\substack{\text{Ferric oxide (red)} \\ Fe^{+3}}}{Fe_2O_3} \xrightarrow{-O_2 \text{ (reduction)}} \underset{\substack{\text{Ferrous oxide (black)} \\ Fe^{+2}}}{2FeO} \xrightarrow{+O_2 \text{ (oxidation)}} \underset{\substack{\text{Ferric oxide (red)} \\ Fe^{+3}}}{Fe_2O_3}$$

The specific soil weathering processes mentioned here operate in conjunction with a multitude of other examples at various levels of intensity on an endless range of rocks and minerals under a great variety of climatic and hydrological conditions. Is it any wonder that such a complex medium as soil results from these processes? Added to that complexity are the many ways that human intervention can alter the environment with which the soil interacts. The full impact of that relationship has yet to be understood.

Human Activities and Soil Formation

Human impact on the soil has increased with the multiplication and diffusion of people on the globe, and, especially, with their technological advances. In the course of trying to provide a favorable environment for life, human populations can knowingly, or unknowingly, manipulate conditions to the extent that they affect soil formation. Human activities act as external modifiers to the internal soil formation processes, such as chemical weathering. The distinction between whether the "land shapes the people" or the "people shape the land" is becoming less clear. Table 7-2 suggests some human activities that can influence soil development, although the rate and extent of influence are certainly open for discussion.

Table 7-2. Human activities can sufficiently impact the other five soil-forming factors so as to be considered a sixth factor. (From: Bidwell and Hole, 1965. Soil Science Vol. 99:65–72.)

Soil-Forming Factor	*Human Activities*	
	Beneficial Effects	*Detrimental Effects*
Parent Material	Fertilization; liming; importing/exporting soil	Nutrient depletion; waste dumping; flooding
Topography	Erosion control; land shaping; landscape structures	Accelerating erosion; excavation; strip mining
Climate	Irrigation; drainage; mulching; wind abatement	Surface clearing; smog; compaction
Organisms	Controlling plant and microbe populations; adding organic matter; controlling disease	Reducing biological diversity; fostering organic matter loss; pollution
Time	Soil amendments; excavation and exposure of subsoil; land reclamation	Accelerated acidification; burying soil under fill, water, or concrete

Experimental Procedure

Materials

- *Fluorapatite, $Ca_5(PO_4)_3F$*
- *Mortar, pestle, 20-mesh and 60-mesh sieves*
- *pH meter or pH indicator paper*
- *Sucrose*
- *Drinking straws*
- *Soil: acid soil, neutral soil, red soil*
- *Ammonium molybdate: Dissolve 5 g $(NH_4)_6Mo_7O_{24} \cdot 4H_2O$ in 50 mL deionized water. Heat and filter if turbid, then add 50 mL concentrated nitric acid and 100 mL deionized water.*
- *Stannous chloride: For stock solution, dissolve 10 g $SnCl_3 \cdot 2H_2O$ in 25 mL concentrated hydrochloric acid and store in a brown glass bottle. Then, each day make a fresh solution by combining 3 mL stock solution and 97 mL deionized water.*
- *Phenolphthalein indicator: dissolve 0.5 g indicator in 800 mL ethyl alcohol and bring to 1000 mL with deionized water.*
- *Chlorophenol red indicator: dissolve 0.4 g indicator in 10 mL ethanol and bring to 1000 mL with deionized water.*
- *Solutions: dissolve the following amount of chemical in deionized water. Bring to 1000 mL volume.*

    0.1 N hydrochloric acid: 8.3 mL concentration HCl
    1.0 N ammonium hydroxide: 35 g NH_4OH
    Saturated ammonium oxalate: 50 g $(NH_4)_2C_2O_4 \cdot H_2O$
    Saturated calcium hydroxide: add $Ca(OH)_2$ until excess solid is evident
    1 M sodium carbonate: 86 g Na_2CO_3
    1 M sodium chloride: 58.4 g NaCl
    1 M calcium chloride: 147 g $CaCl_2 \cdot 2H_2O$
    1 M aluminum chloride: 241.5 g $AlCl_3 \cdot 6H_2O$

Part I. The effect of particle size on rate of weathering

Coarse and fine samples of fluorapatite, $Ca_5(PO_4)_3F$, a common calcium phosphate soil mineral, are "weathered" at an accelerated rate in hot water. The release of calcium and phosphorus are compared to illustrate that fine samples weather faster than coarse samples because they expose more specific surface area to their environment.

1. Weigh 0.5 g of coarsely ground apatite (>20 mesh) and 0.5 g of finely ground apatite (<60 mesh) into two, separate, clean 125-mL flasks.

2. Add 50 mL deionized water and 10 drops of phenolphthalein to each flask.

    Phenolphthalein indicator is colorless in acid or neutral solutions and pink in basic solutions.

3. Place flasks on a hot plate and boil the mixture gently for two minutes. Note the appearance of any pink color and record your results.

Pink color development indicates the hydrolysis of calcium in apatite and the release of $Ca(OH)_2$, which makes the water basic. The more intense the pink color, the greater the extent of hydrolysis.

4. Test each sample for the release of phosphorus as follows:

    a. Filter the solution from each flask and transfer 5 mL of filtrate into separate test tubes.

    b. Add 5 drops of $(NH_4)_6Mo_7O_{24} \cdot 4H_2O$ (ammonium molybdate) solution to each test tube and mix.

    c. Add 1 drop of $SnCl_3 \cdot 2H_2O$ (stannous chloride) solution to each test tube and mix. Compare the blue color intensity after one minute and record your results.

    The intensity of blue color development is directly correlated with the level of phosphorus dissolved from apatite.

Part II. The release of nutrients from soil by weathering

An acid soil and a neutral soil are "weathered" at an accelerated rate in a hot, acid solution. An acid soil indicates a more highly weathered, less fertile condition than a neutral soil and should release fewer nutrients.

1. Weigh 10 g of an acid soil and 10 g of a neutral soil into separate, clean 125-mL flasks.

2. Add 50 mL of 0.1 N HCl (hydrochloric acid) and mix by swirling. Place flasks on a hot plate and boil the mixture gently for two minutes. The acid and heat stimulates and hastens the natural weathering processes.

3. Filter and collect the filtrates in clean beakers.

4. Test each sample for the release of calcium as follows:

    a. Transfer 10 mL of each filtrate into separate test tubes.

    b. Add 5 drops of 1.0 N NH_4OH (ammonium hydroxide) to each test tube and mix.

    c. Add 5 drops of saturated $(NH_4)_2C_2O_4 \cdot H_2O$ (ammonium oxalate) to each test tube and mix. Note the appearance of calcium oxalate, the white precipitate, and record your results.

    The amount of calcium released is determined by the amount of precipitate. To compare precipitate amounts, shake each tube and hold both in front of a black line on a white page. The greater the amount of precipitate in the tube, the more the black line will be obscured.

5. Test each sample for the release of phosphorus as follows:

    a. Transfer 5 mL of filtrate into separate test tubes.

    b. Add 5 drops of $(NH_4)_6Mo_7O_{24} \cdot 4H_2O$ (ammonium molybdate) solution to each test tube and mix.

    c. Add 1 drop of $SnCl_3 \cdot 2H_2O$ (stannous chloride) solution to each test tube and mix. Compare the blue color intensity after one minute and record your results.

    The intensity of blue color development indicates the level of phosphorus dissolved from apatite.

Environmental Influences on Soil Formation

Part III. Examples of carbonation

During normal growth processes, plant roots release carbon dioxide, CO_2, through respiration. Soil microbes also release carbon dioxide during respiration. These CO_2 sources acidify the soil solution because they contribute to the formation of carbonic acid.

1. Fill a 125 mL flask half full with deionized water. Add 3 drops of chlorophenol red.

    Chlorophenol red indicator is red in neutral solutions and yellow in acid solutions.

2. Use a drinking straw and blow your breath into the water for at least one minute. The CO_2 from your breath will form carbonic acid in the water and change the color of the indicator. Note and record any color change.

$$CO_2 \; + \; H_2O \; \longleftrightarrow \; \underset{\text{Carbonic acid}}{H_2CO_3}$$

3. For a second demonstration of carbonation, transfer 25 mL of saturated $Ca(OH)_2$ in a clean, small beaker.

4. Use a drinking straw and blow your breath into the water for at least one minute. The CO_2 from your breath should cause a cloudy precipitate of calcium carbonate to appear. Note and record any changes to the solution.

$$2H_2CO_3 \; + \; \underset{\text{Calcium hydroxide}}{Ca(OH)_2} \; \rightarrow \; \underset{\text{Calcium bicarbonate}}{Ca(HCO_3)_2} \; + \; 2H_2O$$

$$Ca(HCO_3)_2 \; + \; Ca(OH)_2 \; + \; H_2O \; \rightarrow \; \underset{\text{Calcium carbonate}}{2CaCO_3} \; + \; 3H_2O$$

Part IV. Hydrolysis of common soil salts

Hydrolysis means "reaction with water," a process that can turn the aqueous environment acid, basic, or neutral. Salts produce an acid solution if the cation from the salt and the OH^- anion from the water molecule combine strongly and leave excess H^+ cations to lower the pH.

$$Fe^{3+} \; + \; 3H_2O \; \rightarrow \; Fe(OH)_3 \; + \; 3H^+$$

Salts containing Al^{3+} typically produce an acid environment while those containing Na^+, K^+, Mg^{2+} or Ca^{2+} typically produce neutral or basic solutions because these cations do not combine as strongly with OH^- anions.

1. Transfer 10 mL of 1 M Na_2CO_3, $NaCl$, $CaCl_2$, and $AlCl_3$ to separate, small beakers, plastic cups, or test tubes and determine their pH using a pH meter or indicator paper. Record your observations.

Part V. Microbial reduction of iron in anaerobic soils

Actively growing roots and microbes can reduce the O_2 content of soil air faster than it can be replaced by diffusion, creating an anaerobic environment. Such anaerobic conditions can significantly alter soil processes. Low soil oxygen slows organic matter decomposition and, through its effects on redox potentials, can alter the form of inorganic compounds. A visual example of such a change can be seen when soils high in ferric iron (red) turn gray as the iron is reduced to its ferrous form.

Anaerobic soil microorganisms, if supplied with a source of energy, will reduce ferric (Fe^{3+}) iron oxide to ferrous (Fe^{2+}) iron oxide by adding an electron to the iron atom. The reduced ferrous iron form is readily soluble and black-colored, but at high concentrations it can be toxic to plants.

$$Fe_2O_3 \quad \xrightarrow{-O_2 \text{ (reduction)}} \quad FeO$$
Ferric oxide (red)   *Ferrous oxide (black)*

1. Place 200 g of a red soil into two separate jars. To one jar add 10 g sucrose and mix well.

2. Add sufficient deionized water to each jar to cover the soil to a depth of at least 5 cm. Ensure that the jars containing soil will be anaerobic by closing them with a one-hole stopper containing a bent glass tube extended into another jar of water. This permits gases to escape but oxygen cannot enter. The jar containing water should have a two-hole stopper.

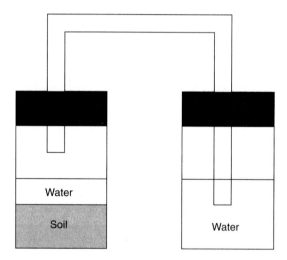

3. Incubate the jars at room temperature and observe color changes in the soil on a weekly schedule. Noticeable color changes should occur within two weeks, but continue the process longer if possible. Record your observations.

Environmental Influences on Soil Formation

Name _____ Section _____

Data

Part I. The effect of particle size on rate of weathering

Particle size	Relative calcium hydrolysis (Pink color intensity)	Relative phosphorus release (Blue color intensity)
Coarse		
Fine		

Part II. Release of nutrients from soil by weathering

Soil type	Relative calcium release (Amount of precipitate formed)	Relative phosphorus release (Blue color intensity)
Acid soil		
Neutral soil		

Part III. Examples of carbonation

Solution	Observations
Water + chlorophenol red	
Saturated $Ca(OH)_2$	

Environmental Influences on Soil Formation

Name _____ Section _____

Data

Part IV. Hydrolysis of common soil salts

Salt	pH	Relative hydrolysis
Na_2CO_3		
NaCl		
$CaCl_2$		
$AlCl_3$		

Part V. Microbial reduction of iron in anaerobic soils

Incubation time	Soil + sucrose		Soil	
	Soil color	Iron form	Soil color	Iron form
Week 1				
Week 2				
Week 3				
Week 4				

Environmental Influences on Soil Formation

Name _____ Section _____

Questions

1. For Part I, describe the significance of the pink color and blue color development in the heated mixture.

2. Describe the chemical changes occurring to the apatite mineral as it undergoes hydrolysis. Why were there differences in the amount of calcium and phosphorus released from the two samples?

3. Which segment of Part I illustrates physical weathering and which segment illustrates chemical weathering?

4. In Part II, how do the differences in the amount of calcium and phosphorus released relate to the rate of weathering of the two soils? Which soil would be considered most fertile? Why?

5. In Part III, how did your breath cause the pH of the water to change? How does carbonic acid get into the soil environment?

Environmental Influences on Soil Formation

Name _____ Section _____

Questions

6. How do you account for the differences in pH between the various salts used in Part IV? What is the significance of hydrolysis in the soil environment?

7. What color changes occurred in the two flasks incubated in Part V? What chemical changes occurred to the iron in these soils?

8. In Part V, what was the role of sucrose in causing the soil to change color? Describe a natural situation where this same effect might be observed.

9. Select two human activities listed as beneficial in Table 7-2 and explain how they might influence soil development.

10. Select two human activities listed as detrimental in Table 7-2 and explain how they might influence soil development.

Exercise 8: Properties of Colloids in the Soil Environment

> The adsorptive capacity of a soil resides with the finest soil particles, clay and humus, known collectively as colloids. The ability of colloids to adsorb and exchange materials lends special chemical, physical, and environmental behavior to soils. Fertilization adjusts the distribution of adsorbed ions to increase crop growth efficiency. Pollution results when the distribution has been altered in ways that make normal use of the soil hazardous.
>
> Exercise Goal: You will demonstrate how soil colloids can adsorb and exchange cations. You will test the ability of various cations to flocculate clays and measure the water adsorptive capacity and swelling factor of two clays.

Most chemicals released into the environment react with soil. The nature of these chemical interactions determine their ultimate environmental impact. Soluble chemicals disperse throughout the soil where they encounter the chemically active solid fraction, collectively called **colloids,** particles less than 1 μm. Note in Table 8–1 that some clay minerals exceed the size definition of a colloid.

The colloidal fraction of soil consists of extremely small particles categorized as either **clay** (mineral colloids) or **humus** (organic colloids). Despite their small size, colloids exert a major influence on a wide range of soil properties and behavior. Colloids are distinguished by an extremely large **specific surface area** (surface to mass ratio) and a net negative electrical charge, which makes them **anionic.** Together, these two properties, high specific surface area and electronegativity, make colloids highly reactive and adsorptive. Remember to distinguish adsorption, where molecules adhere to a surface, from absorption, where molecules cross a membrane.

Colloids can adsorb uncharged molecules, such as water and some pesticides, or positively charged ions and molecules, termed **cations.** Colloids repel negatively charged particles. The activity of a chemical may be enhanced, diminished, prolonged, or significantly changed once it becomes adsorbed. The amount and nature of the materials adsorbed may also change colloidal properties, for example, their acidity or their tendency to aggregate or disperse. This unique interaction between an adsorbed material and soil colloids eventually determines the impact of any chemical in the soil environment.

Colloids vary in physical and chemical characteristics and, hence, vary in their impact on soil properties. Colloids are either **crystalline** (with a highly organized structure) or **amorphous** (without crystallinity). Crystalline types are subclassified by their ratio of tetrahedral and octahedral sheets and whether they shrink and swell upon drying and wetting. Table 8-1 illustrates these differences for three typical clays.

The quantity of cations a mass of colloids can adsorb is called **cation exchange capacity** (CEC) and is expressed as $cmol_c$ kg^{-1} (meq /100 g). Clay and humus maintain a swarm of cations near their surfaces to balance the negative charge carried by the colloid. These cations are adsorbed strongly enough to resist leaching, yet are available for plant uptake or replacement by other cations in the solution.

The cation exchange property of colloids describes a most significant feature, their ability to be a temporary storehouse for cationic plant nutrients (calcium, magnesium, potassium, etc.). Under natural

Properties of Colloids in the Soil Environment

conditions when exchangeable calcium, magnesium, and potassium are removed from the clay and humus by leaching or plant absorption, they are replaced by hydrogen. This exchangeable hydrogen is responsible for soil acidity. Adjustment of the ion composition on colloids is the basis for soil chemistry and fertility management.

Table 8-1. Properties of soil colloids.

Properties	Mineral Colloids			Organic Colloid
	Kaolinite	Illite	Montmorillonite	Humus
Particle diameter, μm	0.1 – 4	0.1 – 2	0.01 – 1	Very small
Particle thickness, μm	0.05	0.005 – 0.03	0.001 – 0.01	Very small
Specific surface area, $cm^2\ g^{-1}$	5 – 20	50 – 300	700 – 800	Very large
Cation exchange capacity, $cmol_c\ kg^{-1}$	3 – 15	15 – 40	80 – 100	~200 (variable)
Maximum water adsorptivity, $g\ g^{-1}$	0.3–0.4	0.3–0.6	6–7	up to 20
Tetrahedral to octahedral sheet ratio	1:1	2:1	2:1	NA
Shrink and swell capability	No	No	Yes	No
Climate prevalence	Subtropical	Temperate	Temperate	Cool, moist

Organic colloids are more adsorptive than mineral colloids are. Humus can adsorb four to five times more water than silicate clays can. Organic colloids also make a significant contribution to a soil's cation exchange capacity because they have considerably larger CEC than do mineral colloids. Typically, the CEC of soil humus is 200 $cmol_c\ kg^{-1}$, but this capacity changes with soil pH. In acid soils, H^+ and Al^{3+} may be bound so tightly as to block more than half of the exchange sites otherwise available to cations at higher pH.

Five properties illustrating the interaction of soil colloids and chemicals in the environment will be studied in this exercise: cation exchange, flocculation, dispersion, shrinking, and swelling.

Cation Exchange

Cations adsorbed on colloidal particles are called exchangeable cations because they can be replaced by, or exchanged for, other cations from the surrounding solution. **Cation exchange** is the process whereby cations replace others already on colloidal sites. The **exchange complex** refers to all soil colloids, both mineral and organic, that adsorb and exchange cations. Soil management practices have been developed that use the cation exchange process to make soils more productive. Fertilizing soils, applying gypsum to sodic soils, and liming acid soils are three such beneficial examples. These processes each induce cation exchange reactions, which change the kind and proportion of adsorbed cations and thereby modify soil properties.

Soil fertility management utilizes soil tests and fertilizer additions to load the cation exchange complex with a desirable distribution of plant nutrients. For example, when potassium fertilizer is added to a potassium-deficient soil, the fertilizer dissolves and potassium ions are attracted to the exchange complex. The amount of exchangeable potassium will be greatly increased, which in turn raises that amount in the soil solution, though only slightly in comparison to that adsorbed (see Fig. 8-1).

Properties of Colloids in the Soil Environment

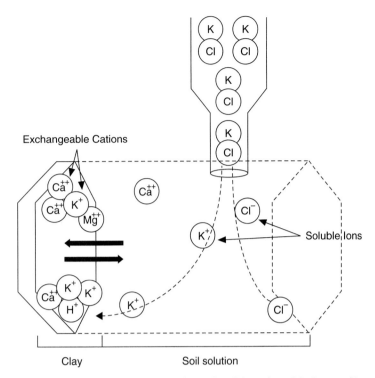

Figure 8-1. The cation exchange process when potassium fertilizer is added to soil.

Liming an acid soil results in basic calcium ions displacing acid hydrogen and aluminum ions from the exchange complex so they can subsequently be neutralized. (Exercise 10 will show how the hydrogen and aluminum are neutralized).

In sodic soils the exchange complex contains excessive sodium, a condition that destroys soil structure and leads to impaired air and water movement. Gypsum ($CaSO_4$) is added to reclaim sodic soils. Calcium ions from gypsum replace the sodium by the cation exchange process. Displaced sodium can then be removed by leaching. The resulting calcium-dominated colloids improve the soil's structure and thereby improve air and water distribution (See Exercise 11).

Plant roots also possess sites that exhibit cation exchange. As plant cells accumulate the cationic nutrients necessary for growth they often release hydrogen to maintain electroneutrality. The hydrogen ions can occupy the sites formerly used by plant nutrients (see Fig. 8-2). The continuation of this process feeds the plant but leaves the soil infertile and acidic and a candidate for fertilization and liming.

Since nutrients in the form of nitrate-nitrogen (NO_3^-) and phosphorus ($H_2PO_4^-$) exist in the soil as anions, they are not found among the exchangeable ions on soil colloids. However, soil nitrogen present as the ammonium cation (NH_4^+) will reside on the exchange complex.

Flocculation and Dispersion of Colloids

The distribution of cations on colloids controls flocculation and dispersion, two colloidal conditions that significantly affect a soil's physical condition. Flocculation promotes aggregation, which in turn fosters good permeability for air, water, and roots. Dispersed soils, on the other hand, are typified by tight,

Properties of Colloids in the Soil Environment

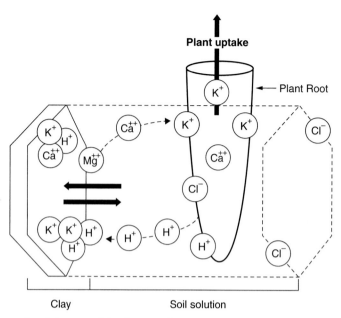

Figure 8-2. The exchange of calcium and hydrogen between soil colloid and plant root.

compacted conditions that result in poor plant growth and create environmental hazards by accelerating runoff and erosion.

Flocculated colloids bind with each other and other soil particles to form porous aggregates. **Dispersed** colloids maintain a separated, unaggregated state conducive to low porosity and low permeability. The desired condition in soils is for colloids, both clays and humus, to promote flocculation. Flocculated colloids foster aggregate development and desirable soil structure.

Whether colloids exist in a flocculated or dispersed condition depends on the type and distribution of cations on their exchange sites. Cations that promote flocculation, including aluminum, calcium, magnesium and hydrogen, possess a common property of high charge density. Charge density is determined by the number of charges a cation possesses and the hydrated volume of the cation (actual size including any attached layers of water). Polyvalent ions with little hydration have high charge density. Highly hydrated, monovalent ions cause dispersion because at typical soil concentrations their charge density is too low to flocculate colloids.

Flocculation and dispersion are complex processes. The net negative charge on colloids would normally cause them to repel each other, but ions with a sufficiently high charge density can neutralize this repulsive effect and allow flocculation. Sodium cations, being monovalent and highly hydrated, have a low charge density and are only weakly adsorbed by colloids. Sodium and other dispersing cations fail to completely neutralize the negative charges, allowing the colloids to repel each other and maintain a dispersed state. Both flocculation and dispersion are reversible by changing the kind and amount of adsorbed cations.

Shrinking and Swelling

Soils containing colloids that shrink and swell will have cracks that appear when dry and disappear after wetting. This soil behavior can contribute to environmental and engineering hazards. Open cracks allow ready entry for materials and facilitate their deep penetration into the soil. A pollutant spill on a cracked soil would increase the contamination zone and complicate remediation. Shrink and swell soils are especially unsuitable for liquid containment sites if drying ever occurs. Soil movement associated with shrinking and swelling can also damage buildings. Structures containing hazardous substances require special, and often costly, modification if they are built on soils susceptible to shrinking and swelling.

The amount of shrinking and swelling exhibited by a soil depends upon its texture (amount of clay), the type of clays present, and to some extent, the type of cations adsorbed on the clays. Montmorillonite clays have an expanding crystal lattice (Table 8-1) and cause soils to shrink upon drying and swell when wet. As the content of sand and organic matter increases, this effect will be reduced. Also, clays dominated by monovalent cations will swell more when wetted than clays dominated by divalent cations.

Clays swell when water enters between the structural layers (lattices) and forces them apart. As evaporation or plant water use dries the soil, this water moves out of the lattices and the clay structure contracts.

Properties of Colloids in the Soil Environment

Experimental Procedure

Four experiments are presented here to illustrate the properties of colloids. Each consists of a hypothesis and an experimental procedure designed to furnish evidence necessary to evaluate the hypothesis. Results of each experiment are designed to be more comparative than quantitative. Use the experimental findings to answer the questions that follow.

Materials

- *Funnels, funnel rack, test tubes, test tube rack, medium-speed filter paper (e.g., Whatman no. 2)*
- *Light-colored, sandy soil; dark-colored, sandy soil; light-colored, clayey soil; dark-colored, clayey soil; acid soil; and neutral soil.*
- *0.2% bentonite suspension: slowly sift 2 g bentonite into 1000 mL deionized water while vigorously stirring.*
- *Bentonite and kaolinite clay in dry powder form (use Wyoming bentonite, if available).*
- *Petri dish, 60 x 15 mm*
- *500 mL plastic beakers, spatula*
- *Balance*
- *Solutions: dissolve the following amount of chemical in deionized water. Bring to 1000 mL volume*
    - *0.02 N barium acetate: 2.55 g $Ba(C_2H_3O_2)_2$*
    - *Saturated potassium dichromate: 100 g $K_2Cr_2O_7$*
    - *0.2 N potassium chloride: 14.9 g KCl*
    - *Saturated ammonium oxalate: 50 g $(NH_4)_2C_2O_4 \cdot 2H_2O$*
    - *0.1 N sodium chloride: 5.84 g NaCl*
    - *0.1 N potassium chloride: 7.46 g KCl*
    - *0.1 N hydrochloric acid: 8.3 mL concentrated HCl*
    - *0.1 N calcium chloride: 7.35 g $CaCl_2 \cdot 2H_2O$*
    - *0.1 N magnesium chloride: 10.2 g $MgCl_2 \cdot 6H_2O$*
    - *0.1 N aluminum chloride: 8.05 g $AlCl_3 \cdot 6H_2O$*
    - *dilute benzyltrimethylammonium chloride: 5 g $C_6H_5CH_2N(CH_3)_3Cl$*

Part I. Cation Adsorption by Colloids

Hypothesis: Soils adsorb cations from the surrounding solution, and those soils with high clay and/or humus content adsorb more cations than do soils with low clay and/or humus content.

In this test a solution containing barium cations is passed through soil. Barium is not normally found in soil, so the solution should be the sole source of this ion. Barium's behavior is representative of any positively charged ion in the environment. If the soil adsorbs barium from the solution, the barium concentration of the solution will be lowered as it passes through the soil. Soils of different clay and humus content are tested to illustrate their difference in cation adsorption capacity.

1. Suspend 4 funnels on a rack and place a test tube beneath each funnel stem. Position a medium-speed filter paper in each funnel.

2. Place 15 g of soil as designated below into each funnel and label:

   Funnel #1: light-colored, sandy soil (low in organic colloids, low in mineral colloids)

   Funnel #2: dark-colored, sandy soil (high in organic colloids, low in mineral colloids)

   Funnel #3: light-colored, clayey soil (low in organic colloids, high in mineral colloids)

   Funnel #4: dark-colored, clayey soil (high in organic colloids, high in mineral colloids)

3. Form a slight depression in the center of the soil in the funnels. Slowly pour 20 mL of 0.02 N $Ba(C_2H_3O_2)_2$ (barium acetate) directly into this depression. Do not overflow the depression as this will allow some of the solution to drain through the filter paper without going through the soil.

4. Collect as much filtrate as possible, then equalize the amounts of filtrate by removing excess from the tubes with the most liquid. To a fifth test tube, add an amount of 0.02 N $Ba(C_2H_3O_2)_2$ to equal that collected from the soils.

5. To each test tube add three drops of saturated $K_2Cr_2O_7$ (potassium dichromate), stopper, and shake. When $K_2Cr_2O_7$ is added to a solution containing barium ions, an insoluble $BaCr_2O_7$ (barium chromate) precipitate forms. The amount of barium ions in the filtrate represents the portion not adsorbed by the soil. The fifth tube shows the amount of precipitate formed when no barium was adsorbed by the soil.

$$Ba^{2+} + K_2Cr_2O_7 \rightarrow 2\,K^+ + BaCr_2O_7\,(ppt.)$$

6. After several minutes compare the amount of precipitate in each tube by shaking all tubes and holding them in front of a dark line on a white paper. The relative amount of precipitate can be judged by how clearly the dark line can be seen. Rank the amount of precipitate present, record your results, and answer the questions for this part.

Part II. Cation exchange

Hypothesis: Cations adsorbed on soil colloids can be replaced by cations added to the soil solution.

In this test, a solution containing potassium cations will be added to soil, and the amount of calcium displaced into the filtrate will be tested. Calcium is selected for testing because it is usually the most prevalent cation in soils. The procedure is designed to evaluate whether the calcium might have originated from a soluble source rather than from the exchange complex.

1. Suspend 4 funnels on a rack and place a test tube beneath each funnel stem. Position a medium-speed filter paper in each funnel.

2. Place 15 g of each of these soils into a funnel and label:

   Funnel #1 and #2: acid soil

   Funnel #3 and #4: neutral soil

3. Form the soil in the funnels with a slight depression in the center. To funnels #1 and #3, slowly add 30 mL of 0.2 N KCl (potassium chloride) directly into this depression. Do not overflow the depression as this will allow some of the solution to drain through the filter paper without going through the soil. The potassium ions (K$^+$) will displace many of the cations already adsorbed on soil colloids into the filtrate being collected in the test tubes.

4. To funnels #2 and #4, slowly add 30 mL of deionized water into the depression. Deionized water contains no cations so it only extracts cations already in the soil solution or dissolved from soluble compounds in the soil, not those on the exchange complex.

5. Collect as much filtrate as possible, then equalize the amounts of filtrate by removing the excess from tubes with the most liquid.

6. Test for calcium in the filtrate by adding 5 drops of saturated $(NH_4)_2C_2O_4$ (ammonium oxalate) to each tube. A cloudy white precipitate of CaC_2O_4 (calcium oxalate) indicates the presence of calcium.

$$Ca^{2+} + (NH_4)_2C_2O_4 \rightarrow 2NH_4^+ + CaC_2O_4(ppt.)$$

7. After several minutes compare the amount of precipitate in each tube by shaking all tubes and holding them in front of a dark line on a white paper. The tube with the most precipitate present will obscure the dark line the most. Rank the amount of precipitate, record your results, and answer the questions for this part.

Part III. Flocculation and dispersion

Hypothesis: Common soil cations with high charge density can flocculate clays, while cations with low charge density cannot. Organic matter in soils also possesses flocculating capabilities.

Wyoming bentonite occurs naturally as a sodium-saturated clay and makes a suspension when dispersed in water, which works quite well for this experiment. When other cations, those commonly found in soil, are added to the suspension, they will exchange with the sodium on the colloid. Flocculation may or may not take place depending upon the cations added.

1. Fill 8 clean test tubes with equal amounts (approximately 3/4 full) of 0.2% bentonite clay dispersed suspension. Label these tubes #1 through #8.

2. Add nothing to tube #1; use it as an untreated reference. Add 10 drops 0.1 N NaCl (sodium chloride), 0.1 N KCl (potassium chloride), 0.1 N HCl (hydrochloric acid), 0.1 N $CaCl_2$ (calcium chloride), 0.1 N $MgCl_2$ (magnesium chloride), and 0.1 N $AlCl_3$ (aluminum chloride) to tubes #2 through #7, respectively. To tube #8 add 10 drops of dilute benzyltrimethylammonium chloride, a cationic organic molecule that strikingly simulates soil organic matter.

3. Stopper and shake the tubes to mix ingredients.

4. Place tubes in a rack and observe the relative floccule formation rate, floccule size, and floccule settling rates. To detect the smallest floccules, hold test tube in front of a light source and compare to the reference tube. Make a final observation of each tube at the end of the laboratory period, then rank the amount of precipitate, record your results, and answer the questions for this part.

Part IV. Shrinking and swelling

Hypothesis: Shrink and swell behavior is related to clay type. Kaolinite clays do not swell upon wetting, while bentonite clays swell to many times their original volume upon wetting.

1. Weigh an empty 400-mL plastic beaker and spatula.

2. Fill a small (60 x 15 mm) petri dish level full with dry bentonite clay (a shrink and swell, montmorillonitic-type clay). Transfer this clay into the plastic beaker and reweigh clay, spatula, and beaker to determine the amount of clay. Save the petri dish.

3. Add water slowly to the dry clay, and mix thoroughly with the spatula until the "liquid limit" is reached. The liquid limit is defined as the moisture content when the soil takes on the fluid properties of a liquid. Bentonite clay goes through several stages before reaching the liquid limit. Upon initial wetting, the clay becomes sticky and clumps together. Adding more water creates a lumpy paste consistency. The clay cannot reach the liquid limit until the lumps of dry clay are wetted. Keep adding water until the clay takes on the consistency and texture of smooth, soft ice cream. At the liquid limit, swirls created by stirring slowly disappear when the spatula is removed. Ask the instructor to demonstrate this condition.

4. After reaching the liquid limit, reweigh the beaker, spatula, and wet clay. Determine the moisture content at the liquid limit according to the formula on the data sheet.

5. Repack the wet clay into the petri dish. Determine how many times the swollen clay will fill the petri dish and designate this number as the "swelling factor."

6. Label your last full petri dish and air dry it until the next lab period. Then estimate the approximate degree of shrinkage.

7. Repeat the procedure for a nonshrinking, nonswelling clay, such as kaolinite.

Properties of Colloids in the Soil Environment

Name _____ Section _____

Evaluation of Hypotheses

Use specific observations from each experimental procedure to describe how you now know whether or not each hypothesis is true.

I.

II.

III.

IV.

Properties of Colloids in the Soil Environment

Name _____ Section _____

Data

Part I. Cation adsorption

Funnel No.	Organic colloids	Mineral colloids	Relative amount of $BaCr_2O_7$ precipitate	Relative amount of barium retained by soil	Relative cation exchange capacity of soil
1	Low	Low			
2	High	Low			
3	Low	High			
4	High	High			
5	Not leached through any soil		Maximum	None	

Data

Part II. Cation exchange

Funnel No.	Soil acidity	Filtrate	Relative amount of calcium in filtrate	Relative amount of calcium from cation exchange complex
1	Acid	KCl		
2	Acid	Water		
3	Neutral	KCl		
4	Neutral	Water		

Properties of Colloids in the Soil Environment

Name _____ Section _____

Data

Part III. Flocculation and dispersion

Tube No.	Cation	Relative flocculation rate	Relative floccule size	Relative settling rate	Charge density (High or Low)
1	None				
2	Na^+				
3	K^+				
4	H^+				
5	Ca^{2+}				
6	Mg^{2+}				
7	Al^{3+}				
8	Organic				

Data

Part IV. Shrinking and swelling

	Bentonite	Kaolinite
1. Weight of beaker, spatula, & dry clay, g		
2. Weight of beaker & spatula, g		
3. Weight of dry clay, g		
4. Weight of beaker, spatula, & wet clay, g		
5. Weight of water, g		
6. Percent water: (water ÷ dry clay) × 100		
7. Swelling factor: number of dishes filled with wet clay.		

Properties of Colloids in the Soil Environment

Name _____ Section _____

Questions

Part I. Cation absorption

1. Which soil adsorbed the most barium? Which soil adsorbed the least barium? Why?

2. Based on texture, which soil is expected to have the greatest cation exchange capacity? Which soil is expected to have the least? Based on your experimental evidence, which soil had the greatest CEC? Which soil had the lowest CEC?

3. What are some naturally occurring materials found in the following sources that might be adsorbed by soil colloids? Which of these would be considered environmental hazards?

    a. rainwater:

    b. runoff from agricultural land:

    c. sewage effluent water:

    d. industrial waste water:

Properties of Colloids in the Soil Environment

Name _____ Section _____

Questions

4. What is the impact of organic matter on cation adsorption?

Part II. Cation exchange

5. Which soil contained the largest amount of calcium on the exchange complex? How is the amount of calcium on the exchange complex related to soil acidity?

6. If calcium was found in the water leachate, what would this suggest about the chemical form of calcium in the soil?

7. How would the results of this experiment be altered if $MgCl_2$ was used to leach the two soils instead of KCl? Explain.

Part III. Flocculation and dispersion

8. Write a cation exchange reaction depicting the exchange taking place between a sodium-saturated clay and the added calcium during flocculation.

Name _____ Section _____

Questions

9. Consider two hypothetical cations, A^+ and B^{++}. On the basis of your results in Part III, which would be expected to cause the most rapid flocculation when added to a clay suspension? Would there be any exception to your observations?

10. Are the colloids in normal soils in a disperse or flocculated condition? Relate flocculation of colloids to structure of the soil and water movement through the soil.

Part IV. *Shrinking and swelling*

11. How is it possible for a soil sample to contain greater than 100% water at the liquid limit?

12. If a shrinking and swelling factor had been calculated for illite, how would it compare to bentonite and kaolinite?

13. Where did all the water go that was needed to bring the bentonite to the liquid limit?

Exercise 9: Types of Soil Acidity

> *Soil chemistry and biochemistry are highly regulated by soil pH. Hydrogen ions in the soil solution set the pH value and are called the active acidity. Hydrogen ions on exchange sites of clay and organic matter furnish a reserve acidity. The ratio of reserve to active acidity describes a soil's buffer capacity, a property that regulates how much soil pH will change in response to environmental influences.*
>
> *Exercise Goal: In this exercise you will learn how to use a pH meter, utilize a displacing solution to extract exchangeable hydrogen, and apply fundamentals of neutralization reactions. Students are encouraged to review the fundamentals of acid-base chemistry.*

Most soil chemical and biochemical activities are sensitive to the acidity of their environment. Soil pH describes the level of acidity and is subject to change because of a variety of external environmental factors associated with soil use. As soil pH changes, so do the reactivity, plant availability, toxicity, and/or solubility of most compounds. The inherent susceptibility of soil pH to change is determined by a soil's buffering capacity, a property related to the distribution of acid between the solution and exchange complex.

Soil acidity is determined by the types of cations occupying the exchange sites on the clays and humus material. These cations can have a reactivity that is either acidic or basic. Of the more abundant cations found in soils, only H^+ and Al^{3+} are considered acidic. Aluminum is acidic because it produces hydrogen ions when it reacts with water.

$$Al^{+3} + 3H_2O \rightarrow Al(OH)_3 \text{ (insoluble)} + 3H^+$$

Cations dissociate, to a greater or lesser degree, from the exchange complex and establish an equilibrium distribution between the exchange surface and the soil solution. The amount of each cation on the exchange surface influences the amount of each in the solution. In this manner, the amount of exchangeable H^+ and Al^{3+}, termed the **reserve acidity,** determines the level of solution H^+ and Al^{3+}, the **active acidity** (Fig. 9-1). The amount of reserve acidity typically exceeds that in solution by three to four orders of magnitude (1,000 to 10,000 times).

The amount of H^+ in solution, and hence the level of acidity, is expressed using the term **pH**. By definition, pH is the negative logarithm of the H^+ ion concentration when concentration is expressed in equivalents per liter (i.e., $pH = -\log[H^+]$). For soils, a numerical pH value describes the concentration of ionic (also called active, soluble, or dissociated) hydrogen in the soil solution dissociated from the cation exchange complex (see Table 9-1). The greater the hydrogen pool on the exchange complex (relative to the pool of basic cations), the greater the number of H^+ available to dissociate and the more acidic the soil solution becomes. Acid soils are those with a pH less than 7, neutral soils have a pH = 7, and basic (alkaline) soils are those with a pH greater than 7. Note that a numerical expression of soil pH describes only the active acidity.

Types of Soil Acidity

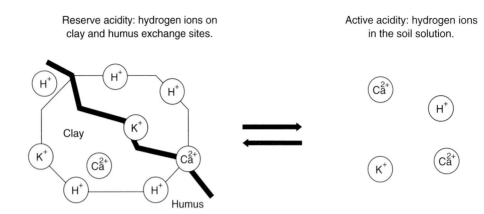

Figure 9-1. Relationship between soil cations and reserve and active acidity.

Table 9-1. Hydrogen ion concentrations at various pH values.

pH	[H$^+$]	pH	[H$^+$]
$-\log$ [H$^+$]	g L^{-1}	$-\log$ [H$^+$]	g L^{-1}
4.8	0.0000160	6.0	**0.00000100**
5.0	**0.0000100**	6.2	0.00000063
5.2	0.0000063	6.4	0.00000040
5.4	0.0000040	6.6	0.00000025
5.6	0.0000025	6.8	0.00000016
5.8	0.0000016	7.0	**0.00000010**

Anything that changes the relative amounts of acidic versus basic cations on the exchange complex of a soil will alter soil pH. Table 9-2 lists some of the more common occurrences that change cation distributions in soils. Optimum soil pH for most plants is approximately 6.8, or slightly acid. At pH 6.8 the exchange complex contains approximately 85% basic cations and 15% acidic cations. If loss of basic cations exceeds their replenishment, dissociation of water provides a small, but inexhaustible, supply of hydrogen to increase acid saturation and lower soil pH.

Table 9-2. Occurrences that alter the acid/base distribution in soils.

	Losses	**Gains**
Bases	Leaching Erosion Plant removal	Fertilizers Liming Manuring Weathering release
Acids	Liming and other neutralization reactions	Fertilizers Dissociation of water Root & microbial activity Aluminum hydrolysis Organic matter decomposition Acid rain

Clay and organic matter provide an ability for soils to **buffer,** or resist, pH change. Most acid (H^+) added to the soil solution will be adsorbed onto exchange sites, a process that converts active acidity into reserve acidity. Concurrently basic cations displaced by H^+ enter the solution. The net result is only a slight change in pH. The buffering reaction works similarly when basic compounds like limestone, $CaCO_3$, are added to soil. As lime neutralizes the active H^+, reserve H^+ dissociates from the exchange sites to prevent a rapid change in soil pH. The **buffering capacity** of a soil is a function of its cation exchange capacity, so those soils with the highest clay and organic matter content are also buffered most effectively. Two soils may have the same pH (active acidity level) but the one with the largest cation exchange capacity will have the greatest amount of reserve acidity and hence the greatest buffering capacity. For this reason it would take less lime to raise the pH of a light-colored, sandy soil from 5.0 to 7.0 than it takes to increase the pH of a dark-colored, clayey soil the same amount.

Neutralization Chemistry

The reserve acidity in a soil is determined by a neutralization reaction. The reaction of an acid (acetic acid, $H^+ C_2H_3O_2^-$) with a base (sodium hydroxide, Na^+OH^-) typifies a neutralization reaction and yields a salt (sodium acetate, $NaC_2H_3O_2$) and water.

$$\underset{acid}{HC_2H_3O_2} + \underset{base}{NaOH} \rightarrow \underset{salt}{NaC_2H_3O_2} + \underset{water}{H_2O}$$

In a neutralization reaction chemically equivalent amounts of acid and base react. One equivalent of sodium hydroxide will neutralize exactly one equivalent of acetic acid (Note: 1 mole of Na = 1 equivalent, 1 mole of Ca^{2+} = 2 equivalents, and 1 mole of Al^{3+} = 3 equivalents.)

To determine the reserve acidity in soil, the soil is leached with a solution of barium acetate. The barium ions will replace the exchangeable hydrogen ions adsorbed on the clay and humus. The replaced hydrogen ions, now being in the soil solution, can be leached out of the soil. The amount of hydrogen ions (acid) leached from the soil is then titrated with sodium hydroxide (a base).

The concentration of the sodium hydroxide solution is expressed in terms of normality. A one normal solution contains one equivalent weight of solute per liter of solution. By measuring the volume of sodium hydroxide solution required to reach the endpoint of the titration, it is possible to calculate the number of milliequivalents of sodium hydroxide added (milliliters of NaOH × normality of NaOH = milliequivalents of NaOH). At the titration endpoint the milliequivalents of sodium hydroxide will equal the milliequivalents of hydrogen ions in the leachate. It is recommended that you complete the supplemental worksheets on soil chemical relationships in the Appendix before proceeding with the experimental procedure.

Types of Soil Acidity

Experimental Procedure

Materials

- Acid soil and neutral soil. (Ideally these two soils should be the same soil type with one sample collected in its native acid condition and the other sample from a field that has been limed and has a reaction near neutral. If these soils are not available, select an acid soil and a soil of similar texture with a neutral reaction.)
- pH meter and standard solutions for pH calibration
- small glass beakers or paper cups
- filtration funnels
- 0.5 N barium acetate : dissolve 63.8 g $Ba(C_2H_3O_2)_2$ in deionized water and bring to 1000 mL volume.
- phenolphthalein indicator: dissolve 0.5 g phenolphthalein in 800 mL ethyl alcohol and bring to a final volume of 1000 mL with deionized water.
- 0.01 N sodium hydroxide: dissolve 0.4 g NaOH in deionized water and bring to 1000 mL volume. Titrate against a standard acid solution to determine exact normality.

Part I. Determination of soil pH (active hydrogen ion content)

1. For soils #1 (acid) and #2 (neutral), mix 10 g soil and 10 mL deionized water in a small beaker or cup. Let stand 10 minutes, stirring occasionally.

2. Test pH of soil-water mixture with a pH meter.

   • A pH electrode senses the difference between the H^+ concentration in the soil solution and in a reference solution inside the electrode. Calibrate the pH meter with reference solutions of a known pH before using.

3. Convert soil pH to H^+ concentration using Table 9-1 or the formula $pH = -\log[H^+]$. Record your results on the data sheet.

Part II. Determination of reserve hydrogen

1. Suspend 2 funnels on a rack and place a clean Erlenmeyer flask under each one. Position a medium-speed filter paper in each funnel (e.g., Whatman No. 2 paper).

2. In one funnel place 10 grams of soil #1 (acid) and in the second funnel place 10 grams of soil #2 (neutral). Form the soil in the funnels with a slight depression in the center.

3. Slowly add 15 mL of 0.5 N $Ba(C_2H_3O_2)_2$ (barium acetate) to each soil directly into the depression to ensure that all filtrates go through the soil. Allow to drain. Barium replaces exchangeable hydrogen, and acetic acid is formed in the solution.

$$\text{Clay} \begin{array}{c} - H^+ \\ - H^+ \end{array} + Ba^{2+} + 2C_2H_3O_2^- \longleftrightarrow \text{Clay} = Ba^{2+} + 2H^+ + 2C_2H_3O_2^-$$

4. Slowly add 30 mL of deionized water in the same manner. Collect both the barium acetate leachate and water wash into the same flask.

5. Titrate the contents of each flask to determine the amount of hydrogen removed from the soil as follows: Add 10 drops of phenolphthalein to the filtrate, then add 0.01 N NaOH dropwise from a buret while continuously swirling the mixture. Continue adding NaOH until the mixture reaches the endpoint (i.e., a permanent pink color). Record the amount of NaOH added to each filtrate.

6. Prepare a blank by mixing 15 mL of 0.5 N $Ba(C_2H_3O_2)_2$ and 30 mL distilled water. Titrate as above and record the volume of NaOH required to reach the endpoint. It will be necessary to subtract the volume of NaOH used to neutralize the blank from the volume used to neutralize each soil's leachate.

7. Record your data, calculate the remainder of the data, and answer the questions on the question sheets.

Types of Soil Acidity

Name _____ Section _____

Data

Active Acidity

Enter the pH of the soils into this table and complete the calculations. For this problem, assume that the furrow slice of a hectare weighs 2,242,000 kg and the moisture content is 22%. Other values of interest: water weighs 1.0 kg L^{-1}, the equivalent weight of H$^+$ = 1 g eq^{-1}, and the equivalent weight of CaCO$_3$ = 50 g eq^{-1}.

Active Acidity	Soil No. 1	Soil No. 2
1. pH		
2. Solution H$^+$ concentration, g L^{-1} Use pH = $-$log [H$^+$] or see Table 9-1		
3. Mass of water, kg ha^{-1}		
4. Volume of water, L ha^{-1}		
5. Mass of H$^+$, g ha^{-1}		
6. Mass of CaCO$_3$ required to neutralize the active H$^+$ per hectare, g		

Types of Soil Acidity

Name _____ Section _____

Data

Reserve Acidity

Reserve Acidity	Soil No. 1	Soil No. 2
1. Weight of soil, g		
2. NaOH added to soil leachate at endpoint, mL		
3. NaOH added to $Ba(C_2H_3O_2)_2$ blank at endpoint, mL		
4. NaOH required for neutralization, mL (Line 2 − Line 3)		
5. Amount of Na^+ added, meq (milliequivalents = mL × N)		
6. Exchangeable H^+ in leachate, meq ($meq_{acid} = meq_{base}$)		
7. Reserve H^+, g g^{-1} soil		
8. Reserve H^+, g ha^{-1}		
9. Buffering ratio, reserve acidity : active acidity (report as ???? : 1)		
10. $CaCO_3$ required to neutralize reserve H^+, kg ha^{-1}		
11. Limestone requirement, lb $acre^{-1}$		

Types of Soil Acidity

Name _____ Section _____

Questions

1. Define the terms "reserve" and "active" acidity. Explain how each is affected by liming a soil.

2. How is the term buffering capacity related to active and reserve acidity?

3. Compare the buffering capacity of the two soils tested. Describe the relationship between buffering capacity of a soil and its cation exchange capacity (CEC).

4. Soil A is a sandy loam and soil B is a clay loam. Both have a pH of 5.2:

    a. Do they have the same amount of active acidity? Explain.

    b. Would you expect them to have the same amount of reserve acidity? Explain.

5. Explain why active pH determinations are so useful in soil studies when at best they measure only a very small portion of the total hydrogen in the soil.

Types of Soil Acidity

Name _____ Section _____

Questions

6. Acid rain deposits about 0.5 lb of H^+ per acre per year on U.S. soils, mostly as H_2SO_4 and HNO_3. How much $CaCO_3$ would be needed (lb per acre) to neutralize this annual acid deposition? (Hint: The equivalent weights of H^+ and Ca^{2+} are 1 g eq^{-1} and 50 g eq^{-1}, respectively.)

7. Some forms of nitrogen fertilizer acidify soil. If 100 lb per acre of ammonia (NH_3) were added and underwent the following reaction, how much acidity, H^+, would be produced per acre each year? (Hints: One NH_3 produces 1 H^+. Ammonia weighs 17 g mol^{-1} and H^+ weighs 1 g mol^{-1}, so the weight of H^+ produced is 1/17th the weight of the ammonia added.)

$$NH_3 \;+\; 2O_2 \;\rightarrow\; H^+ \;+\; NO_3^- \;+\; H_2O \qquad (Atomic\ weights: \;\; N = 14\ and\ H^+ = 1)$$

How much $CaCO_3$ (limestone) should it take to neutralize the H^+ produced from 100 lb of ammonia fertilizer?

(Note: Because some NH_3 is taken up directly as NH_4^+, not all NH_3 undergoes nitrification, so the acidity produced under normal management is only about half of the theoretically possible value. Thus, ammonia fertilizer is not quite as acidifying as it first appears. However, net acidification of soils from ammonium, ammonia, and urea fertilizers and from N fixed by legumes is much greater than that from acid rain.)

8. Why does the acidification from acid rain or nitrogen fertilizer have a greater impact on light-colored, sandy soils than on dark-colored, clayey soils?

Exercise 10: Liming Acid Soils

> *Acidity often results in suboptimal chemical and biological conditions in soil. Both natural conditions and environmental mismanagement can cause soils to become acid. Correcting soil acidity requires application of a liming material. Understanding soil acidity and its correction requires application of principles involving ion activity, buffering capacity, and cation exchange.*
>
> *Exercise Goal: In this exercise you will apply procedures to analyze soil pH, establish a corrective lime requirement, and determine limestone quality. Instructions for modifying a lime requirement based on lime quality and application depth are given.*

Acidity controls important chemical and biological activities in soil and serves as an important criteria for assessing productivity and soil use potential. Nutrient availability and mobility, microbial activity, and herbicide effectiveness are examples of soil processes influenced by soil pH. Soils tend to become acid with time and use, but liming will correct soil acidity and improve pH-related processes.

Soil Acidity

Soil acidity is determined by the proportion of hydrogen (H^+) ions to hydroxyl (OH^-) ions in the soil solution. A soil pH of 7 is neutral, reflecting an equal concentration, 10^{-7} mol L^{-1}, of hydrogen and hydroxyl ions. Soils with pH values below 7 are **acid** (more H^+ than OH^-), and above 7 they are **alkaline** (more OH^- than H^+). Because the pH scale is logarithmic, each one unit change in pH represents a 10-fold change in hydrogen ion concentration. Four pH ranges serve as useful indicators of soil conditions: at pH values below 5.2, toxic levels of exchangeable aluminum and a deficiency of basic nutrient cations (K, Mg, Ca, etc.) are common; pH values in the range 6.5–6.9 are optimum for most plant growth processes and represent a maintenance target for soil management, a pH of 7.8–8.2 is indicative of free calcium carbonate; and a pH greater than 8.5 points to high levels of exchangeable sodium and undesirable sodic conditions.

Soil pH is a measure of hydrogen ion concentration in the soil solution and does not include hydrogen ions held on cation exchange sites of clay and organic matter. Hydrogen ions in the soil solution comprise the **active acidity,** while hydrogen ions adsorbed by the exchange complex make up the **reserve acidity.** The reserve acidity has no direct impact on soil pH, but is in equilibrium with the active acidity and is normally many times more plentiful (see Exercise 9). A large ratio of reserve acidity to active acidity creates an effective buffer against sharp pH fluctuations. Whereas active acidity assesses the need to lime a soil, the reserve acidity determines the amount of liming material necessary to change soil pH.

Reserve acidity depends on several factors affecting the exchange complex in soil, such as the amount and type of clay, organic matter content, and soluble aluminum concentration in the soil. Soils with high clay and/or organic matter content generate a large cation exchange capacity and, consequently, can store a large amount of reserve acidity. Soils with high reserve capacity would require more liming material to change pH than would soils with low reserve capacity. Thus, two soils may have the same soil pH, but show different lime requirements. Aluminum compounds affect soil acidity through reactions that contribute hydrogen ions (H^+) to the soil solution.

Causes of Acid Soils

Soil acidification is a natural process, but some management and environmental situations can accelerate this process. The basic process of soil acidification involves the removal of bases and their replacement with hydrogen ions. High rainfall climates foster low soil pH as water movement through soil leaches basic cations (calcium, magnesium, potassium, and sodium), leaving exchange sites filled with hydrogen ions. Natural rainfall has a pH around 5.6 because it contains carbonic acid formed from reactions of atmospheric CO_2.

Soil microorganisms also contribute to soil acidification through their participation in biochemical reactions that release hydrogen ions. One such reaction is nitrification where ammonium (NH_4^+) is converted to nitrate (NO_3^-) plus hydrogen ions. Fertilizers, or any other soil amendment, containing ammonium or amine nitrogen accelerate soil acidification through nitrification. Plants also contribute to acidification by selectively removing basic cations from the soil to meet growth requirements and replacing them with H^+ from root respiration processes.

Pollutants in the atmosphere can also be sources of soil acidification. Atmospheric oxidation of sulfur and nitrogen oxides form soluble sulfuric and nitric acids that enter soils as acid rain. The pH of acid rain may drop below 4.0. Other airborne, acid-forming compounds from industrial activities can reach soils as dry deposits. The effect of these depositions on soil pH varies with each soil's ability to buffer acid additions.

Plant and Microbial Response to Acid Soils

Each organism has a soil pH range necessary for optimal growth. Some plants thrive in rather acid or alkaline soils, but most prefer a pH near neutrality. Within the optimum pH range, the reactions and processes essential to growth occur at their most ideal rate and toxins in the environment are minimized. Soils too acid for optimum growth of a desired plant type should be limed.

Beneficial organisms tend to dominate soil microbial activities when soil pH is maintained near neutrality. Bacterial activities are most inhibited by acid conditions, while soil fungi are most inhibited by alkaline conditions. Root nodule bacteria responsible for nitrogen fixation in legumes function best near neutrality. Beneficial effects of soil organisms, like organic matter decomposition, a factor in aggregation and associated improvement in soil aeration and drainage, also occur optimally when soil pH is near neutrality.

Environmental Associations to Acid Soils

The behavior of hazardous elements and compounds in the soil environment is sensitive to soil pH. Heavy metals (arsenic, cadmium, lead, mercury, zinc, etc.) typically increase their desorption and solution concentrations in soil under acid conditions. As liming raises the pH of acid soils, metal adsorption onto soil colloids increases, which in turn decreases their mobility and redistribution in nature.

The use of pesticides entails some hazards to the environment and their effectiveness is often related to soil pH. Once pesticides enter soil a major factor governing their fate is adsorption onto colloids. Adsorption decreases their concentration in solution, chemical effectiveness, and mobility in the environment and is related to soil pH in two ways. First, the adsorption capacity on several soil colloids increases as pH increases. Second, pH regulates the degree of protonation (formation of charged groups) of the pesticide molecule and, hence, its capacity for binding to colloids.

Some mining operations involving sulfur-laden ores generate strongly acid wastes when these deposits oxidize to form sulfuric acid. Runoff and erosion of these deposits pose a threat to neighboring land and water. Stabilization of these sites with vegetation requires liming and, often, fertilizer treatments.

How Limestone Raises Soil pH

Limestone is a naturally occurring, soft, sedimentary rock composed predominantly of calcium carbonate ($CaCO_3$), but usually containing some magnesium carbonate ($MgCO_3$) and/or other impurities. If significant amounts of magnesium carbonate are present, the material is called dolomite. Limestone containing both calcium and magnesium is considered a premium material because, in addition to correcting soil acidity, it supplies two plant nutrients instead of just one.

Acid neutralization begins when limestone slowly dissolves in soils, forming calcium (Ca^{2+}) and carbonate (CO_3^{2-}) ions (Fig. 10-1). The calcium ions are attracted to exchange sites where, by mass action, they displace the hydrogen ions (H^+) comprising the reserve acidity. This exchange reaction raises the base saturation on the colloids, a necessary step in decreasing soil acidity. The displaced hydrogen ions join the active acidity in the soil solution where they can react with carbonate ions to temporarily form carbonic acid (H_2CO_3). If carbonic acid was a stable compound, liming would actually increase acidity in the soil solution. But its instability leads to a rapid decomposition to water (H_2O) and carbon dioxide (CO_2). The carbon dioxide molecule, being a gas, escapes from the soil and prevents reformation of carbonic acid. The net effect of the liming process repositions hydrogen ions from where they were causing soil acidity so that they now reside as part of the neutral water molecule. Also, as the liming reaction raises soil pH, aluminum ions are converted to complex aluminum hydroxy ions and lose their ability to generate hydrogen ions by hydrolyzing water.

This neutralization reaction happens slowly in soils because limestone is slow to dissolve and has limited mobility due to its rapid attraction to cation exchange sites. Fine limestone particles will react faster than coarse particles. Also, thoroughly mixing limestone within the soil speeds its reaction. Generally, acid neutralization is maximized one growing season after limestone application.

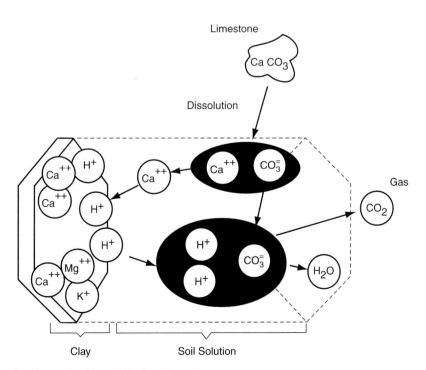

Figure 10-1. Neutralization of soil acidity by limestone.

Liming Acid Soils

Lime Requirement Determination

The lime requirement is the amount of liming material needed to raise soil pH to an optimum range. Making a lime requirement involves three steps:

1. Soil Evaluation

    a. *Measure soil pH (active acidity) to determine the need for liming.*

    b. *Establish the rate of liming needed by measuring the amount of reserve acidity.*

2. Limestone Quality Evaluation

    a. *Determine the calcium carbonate equivalence (CCE) by testing the acid neutralizing capacity, or purity, of the limestone.*

    b. *Determine the fineness factor with sieve analysis.*

    c. *Calculate the effective calcium carbonate rating (ECC).*

3. Lime Rate Adjustment

    a. *Adjust lime rate based on ECC rating.*

    b. *Adjust lime rate based on incorporation depth.*

Materials

- *Small glass beakers or disposable paper cups*
- *Stirring rods*
- *pH meter*
- *Limestone sample*
- *Sieve nest: bottom pan, 60-mesh sieve, 8-mesh sieve, lid*
- *Balance*
- *250 mL beakers*
- *Hot plate*
- *SMP buffer: Dissolve the following materials in deionized water and make to a final volume of 1000 mL. Adjust the pH of the mixture to 7.5 with NaOH.*

    1.8 g para-nitrophenol
    2.5 mL triethanolamine
    3.0 g potassium chromate, K_2CrO_4
    2.0 g calcium acetate, $Ca(C_2H_3O_2)_2$
    53.1 g calcium chloride, $CaCl_2 \cdot 2H_2O$

- *0.5 N HCl: add 41.3 mL concentrated HCl to deionized water and bring final volume to 1000 mL*
- *Phenolphthalein indicator: Dissolve 0.5 g phenolphthalein in 800 mL ethyl alcohol and bring final volume to 1000 mL with deionized water.*
- *0.5 N NaOH: dissolve 20 g NaOH in deionized water and bring final volume to 1000 mL. Titrate against a standard acid solution to determine exact normality.*

Part I. Soil Evaluation

A. Active acidity

Does the soil need liming? Soil pH (active acidity) measures the H^+ concentration in the soil solution and indicates whether a soil should be limed. Generally soils need liming if the pH is 6.4 or lower. Although pH 6.5–6.9 is considered optimum for most plants, economical returns from liming are seldom realized until the pH drops to 6.4. (In some regions, for some acid-tolerant plants, and on some soil, especially organic soils, lime will be recommended only at lower pH values.)

1. For each soil, prepare a 1:1 soil-water mixture in a small beaker or paper cup using 5 g soil and 5 mL deionized water. Stir and let stand for 15 to 20 minutes.

2. Stir mixture again and determine pH on a pH meter. Record this value on the data sheet.

3. If pH is 6.4 or less, this soil needs liming, but another test is needed to determine the rate of liming. Save this mixture for determining the actual limestone requirement in Part I-B.

4. A pH of 6.5 or above indicates no need for liming.

B. Reserve acidity

The amount of lime necessary to counteract soil acidity is determined by the amount of reserve acidity in the soil. Reserve acidity is the H^+ on cation exchange sites of soil colloids, so the reserve capacity depends on the amount and type of clay and the organic matter content. Soils with a large cation exchange capacity have the potential for a large reserve acidity, which in turn would necessitate a large lime requirement for neutralization. Thus, two soils with the same active acidity (pH) can have different lime requirements if they have different cation exchange capacities.

The reserve acidity is determined with a buffered solution analysis. The SMP buffer solution described here is buffered to a pH of 7.5 and contains cations that will replace H^+ on the soil's exchange complex. The pH of the buffer drops in proportion to the amount of acid (H^+) removed from the exchange sites. Table 10-1 correlates the buffer pH to the amount of 100% effective calcium carbonate needed to raise soil pH to an optimum value. (Note: Some states use a buffering solution different that this one. In that event your instructor may want to modify the following procedure and use that reagent.)

1. Add 10 mL of SMP buffer solution to each of the 1:1 soil-water mixtures from Part I-A that have pH 6.4 or lower.

2. Stir intermittently for the next 20 minutes, then determine the pH of the soil-buffer solution on a pH meter.

3. Record that value on the data sheet and use Table 10-1 to determine the lime requirement.

Note: This value is sometimes called "Lime Index" to avoid confusing the buffer pH with the soil solution pH.

Table 10-1. SMP buffer pH and amount of 100% effective calcium carbonate required to raise soil pH to 6.8.

Soil-buffer pH	100% Effective Calcium Carbonate Requirement	
	(kg/ha)	(lb/acre)
Over 7.1	None	None
7.1	1,100	1,000
7.0	1,700	1,500
6.9	2,200	1,800
6.8	2,800	2,500
6.7	3,400	3,000
6.6	4,500	4,000
6.5	5,600	5,000
6.4	6,700	6,000
6.3	8,400	7,500
6.2	9,500	8,500
6.1	10,600	9,400
6.0	12,300	11,000
5.9	13,400	12,000
5.8	14,600	13,000
5.7	15,700	14,000

Part II. Limestone quality evaluation

Limestone quality depends on both its purity and fineness. The limestone rates listed in Table 10-1 are based on pure, finely ground $CaCO_3$. However, since this quality of liming material is seldom available, these rates must be adjusted to compensate for quality of material on hand.

The term *effective calcium carbonate (ECC)* expresses the quality or neutralizing value of a liming material. Two factors determine limestone quality: (1) fineness or particle size, and (2) purity or calcium carbonate equivalency (CCE). In some areas, the purity factor is referred to as the material's neutralizing value (NV), and that term can be used interchangeably with CCE. Quality is expressed by the following formula:

$$ECC = Fineness\ Factor \times CCE$$

The finer the liming material, the higher will be its fineness factor and, hence, the quality of the limestone. Fine particles present a high specific surface area and consequently react faster to neutralize soil acidity. Also, fine material can be mixed more uniformly throughout the soil. On the other hand, some slightly coarse particles are desirable to extend the effectiveness of the liming material over a longer period of time. However, very coarse particles (>8 mesh) react too slowly to contribute significantly toward reducing soil acidity.

The calcium carbonate equivalency (CCE) compares the neutralizing power of the limestone sample with an identical weight of pure $CaCO_3$. Most limestone materials have CCEs less than 100% because they contain natural impurities. Only liming materials with a higher concentration of calcium and/or magnesium than pure $CaCO_3$ can have CCE values greater than 100 (Table 10-2).

Table 10-2. Common liming sources, their composition, and calcium carbonate equivalence.

Liming Material	Composition	CCE Range (%)
Limestone (calcitic)	$CaCO_3$	80–100
Limestone (dolomitic)	$CaCO_3$ and $MgCO_3$	100–108
Marl	$CaCO_3$ with impurities	70–90
Burnt lime	CaO	150–179
Hydrated (slaked) lime	$Ca(OH)_2$	120–136
Industrial lime waste	$CaCO_3$ with impurities	80–100

A. Fineness factor

(Note: Not all states use the same particle size ranges to establish the fineness factor, and your instructor may want to substitute local modifications.)

1. Weigh a 100 g sample of limestone into a beaker.

2. Pour limestone sample onto a sieve nest composed of an 8-mesh sieve, a 60-mesh sieve, and a bottom catch pan. Cover and sieve the sample by shaking vigorously.

3. Weigh and record the amount of material retained by each level of the sieve nest. Save the material finer than 60-mesh for CCE determination.

B. Calcium carbonate equivalency determination

1. Weigh 1.0 g each of limestone (finer than 60 mesh) and laboratory grade $CaCO_3$ into separate 250 mL flasks.

2. Add exactly 50 mL of 0.5 N HCl into each flask and swirl.

3. Gently warm the flasks on a hot plate for 20 minutes. **Do not let sample boil!**

4. Cool flasks to room temperature, add 5 drops of phenolphthalein indicator, and titrate each with 0.5 N NaOH to the endpoint. At the endpoint the indicator turns pink but is difficult to see in the murky solution. Titrate with care.

5. Record normality and volume of NaOH on the data sheet and complete all calculations.

Part III. Limestone rate recommendation adjustment

Information from the soil and limestone evaluations can now be used to establish an adjusted recommendation that will correct soil acidity problems. This final recommendation will take into consideration soil acidity, soil buffering capacity, quality of liming material, and an adjustment, if any, for depth of incorporation. Following this recommendation, a limestone treatment results that, when mixed with soil, will slowly dissolve and gradually raise pH, reaching a maximum approximately one growing season later. Remember, however, that liming an acid soil corrects the accumulated acidity but not the cause of acidity accumulation. Soil pH should be monitored every two to three years to see when re-liming becomes necessary.

A. Adjusting limestone recommendation for quality

The limestone recommendation is calculated by adjusting the amount of 100% effective calcium carbonate required by the SMP buffer test (Part I) with the limestone quality value established in Part II. For example, if the buffer test calls for 3000 kg ha^{-1} and the ECC rating of available limestone is 60%, the limestone recommendation is 5000 kg ha^{-1}.

$$\textit{100\% ECC rate (kg ha}^{-1}\textit{)} \div \textit{ECC rating} = \textit{Limestone recommendation (kg ha}^{-1}\textit{), or}$$

$$\textit{3000 kg ha}^{-1} \div \textit{0.6} = \textit{5000 kg ha}^{-1}$$

This example illustrates that 5000 kg of 60% effective calcium carbonate limestone is required to supply the neutralizing power of a 3000 kg recommendation of 100% effective limestone.

B. Adjusting limestone recommendation for depth of incorporation

The amount of limestone necessary to neutralize soil acidity also depends on the depth to which the material will be incorporated, or the mass of soil to be neutralized. Most recommendations are based on neutralization of about the top 18 cm (~7 inches) of soil. If lime is to be incorporated deeper, more liming material will be necessary and an adjustment to the recommendation is needed (Table 10-3). Conversely, if lime is incorporated shallower than 18 cm, less lime will be needed.

Table 10-3. Factors for adjusting limestone recommendations based on depth of incorporation.

Limestone Incorporation Depth		Rate Adjustment Factor
cm	inches	
8	3	0.4
13	5	0.7
18	7	1.0
23	9	1.3
28	11	1.6

If the 5000 kg ha^{-1} recommendation from the earlier example was going to be incorporated 23 cm deep (9 inches), then the recommendation should be increased by a factor of 1.3, up to 6500 kg ha^{-1}.

Liming Acid Soils

Name _____ Section _____

Data

Part I-A. Soil pH measurement (Does this soil need liming?)

Soil ID	Soil pH	Does this soil need liming?	
		Yes	No
		Yes	No
		Yes	No
		Yes	No

Part I-B. Soil-buffer pH measurement (what is the 100% ECC lime requirement?)

Soil ID	Soil-Buffer pH	Lime Requirement (Table 10-1) 100% ECC, kg ha^{-1}

Part II-A. Limestone fineness factor (Sum of distribution percentage times activity factor)

Sieve Size	Limestone, g (Distribution %)	Activity Factor	Percent Available
> 8 mesh		0.0	
8 – 60 mesh		0.5	
< 60 mesh		1.0	
Fineness Factor (sum) =			

Liming Acid Soils

Name _____ Section _____

Data

Part II-B.   Calcium carbonate equivalence

	Limestone	Pure CaCO$_3$
1. Weight of limestone sample, g	_____	_____
2. HCl acid added, mL	_____	_____
3. HCl normality, N	_____	_____
4. Milliequivalents acid, meq (N × mL)	_____	_____
5. NaOH base added, mL	_____	_____
6. NaOH normality, N	_____	_____
7. Milliequivalents of base, meq (N × mL)	_____	_____
8. Milliequivalents of HCl acid neutralized by NaOH base during titration, meq (Line 7)	_____	_____
9. Milliequivalents of HCl acid neutralized by sample before titration, meq (Line 4 − Line 7)	_____	_____
10. Milliequivalents of CaCO$_3$ in 1 gram, meq (Line 9)	_____	_____ *
* Pure CaCO$_3$ should have 20 meq g^{-1}.		
11. Limestone Calcium Carbonate Equivalence (CCE), % [(Line 10 for Limestone ÷ Line 10 for CaCO$_3$) × 100]	_____	

Part II-C.   Effective calcium carbonate rating

Fineness Factor, %   ×   CCE, %   =   Effective Calcium Carbonate Rating (ECC), %

   (Part II-A)   (Part II-B)

_____ × _____ = _____

Liming Acid Soils

Name _____ Section _____

Data

Part III. Limestone recommendation

Soil ID	Lime Requirement, (100% ECC, Part 1-B) kg ha^{-1}	÷	ECC (Part II) %	×	Depth Factor (Table 10-3)	=	Limestone Recommendation kg ha^{-1} lb acre^{-1}
_____	_____	÷	_____	×	_____	=	_____ _____
_____	_____	÷	_____	×	_____	=	_____ _____
_____	_____	÷	_____	×	_____	=	_____ _____
_____	_____	÷	_____	×	_____	=	_____ _____

Liming Acid Soils

Name _____ Section _____

Questions

1. Why is it necessary to know both the active and reserve acidity before liming soils?

2. Describe how environmental quality might deteriorate if soil acidity is not corrected.

3. How does the liming requirement analysis account for soil differences in cation exchange capacity (i.e., differences in clay and organic matter content)?

4. What are the characteristics of a high-quality limestone?

5. Describe two processes associated with rainfall that contribute toward making soil acid.

6. Write the neutralization process between limestone and an acid soil as a chemical reaction. Also write the reaction with burnt lime substituted for limestone.

Exercise 11: Soil Degradation by Salinity and Sodicity

> *Soils are degraded when too much soluble salt creates salinity, or too much exchangeable sodium causes sodicity. Both conditions adversely affect plant growth. Salinity lowers water availability and sodicity makes soils impermeable to air and water, which in turn fosters runoff and erosion. Mismanagement of irrigated water, especially in dry climates, contributes significantly to these types of soil degradation.*
>
> *Exercise Goal: You will learn about soil tests necessary to identify saline and sodic soils. The decrease in soil permeability caused by sodic conditions is illustrated with a percolation test. Sensitivity of seed germination to soil salinity is also evaluated.*

The chemical environment of the soil solution can become "overloaded" with respect to soluble salts in general, and sodium in particular. At low levels neither soluble salts nor sodium cause any problem, but at elevated levels they create serious limitations to soil use. These undesirable chemical conditions can result from natural occurrences or be induced by human activities. Use of poor quality irrigation water on soils in dry climates generally leads to saline and sodic problems. As increasing demands on the world's fresh water supply reduce both its availability and quality, the soil environment looms as a likely degradation target.

Salinity is a high salt level in the soil solution. Typically, salts formed of the cations calcium (Ca^{2+}), magnesium (Mg^{2+}), and sodium (Na^+) with the anions chloride (Cl^-), sulfate (SO_4^{2-}), and bicarbonate (HCO_3^-) are involved. While these ions are mineral-weathering products commonly found in all soils, at high concentration they create saline conditions. Critical conditions are determined by a soil test that measures the soluble salt concentration in the soil solution.

The soluble salt concentration is determined by measuring the **electrical conductivity** (EC) of the soil solution. Pure water conducts electricity poorly, but EC increases as the content of ions in solution increases. The units used to express EC are deciSiemens per meter. A soil is considered saline if an extracted sample of the soil solution has an EC greater than 4 dS m^{-1} (Table 11-1).

Table 11-1. Characteristics of salt- and sodium-affected soils.

	Normal Soil	Saline Soil	Sodic Soil	Saline-Sodic
pH	< 8.5	< 8.5	> 8.5	< 8.5
Electrical conductivity, dS m^{-1}	< 4	> 4	< 4	> 4
Exchangeable sodium, %	< 15	< 15	> 15	> 15

Salinity causes problems by increasing the osmotic tension of water. This means the ions in solution attract water molecules and force plants to expend additional energy to acquire this bound water. Plants exhibit considerable variability in their tolerance of salinity related to overall soil water conditions, type of salt, type of plant, and stage of development (Table 11-2). Germination is an especially salt-sensitive process.

Table 11-2. Plant response to soil salinity levels.

Salinity effects mostly negligible	Growth of very sensitive plants may be restricted	Growth of many plants restricted	Only tolerant plants grow satisfactorily	Only a few very tolerant plants survive
2	4	8	16	
Electrical Conductivity, dS m^{-1}				

Salts accumulate in soils when their addition exceeds their removal. Water is typically both the source of addition and removal. Buildup begins whenever incoming water evaporates and leaves the salts behind rather than percolating through soils and carrying the salts into groundwater. For this reason, salinity is a constant hazard in arid and semiarid climates. The continued application of irrigation water, even though it may be low in soluble salts, eventually adds more salt than is removed. The use of poor-quality irrigation water (i.e., having a high soluble salt level) accelerates the salt buildup process.

Sodic soils contain sufficient exchangeable sodium to degrade soil physical properties. Sodium (Na^+) is a cation that will be weakly adsorbed to soil cation exchange sites. Though sodium is common to practically all soils, when present in a sufficiently high proportion it creates sodic conditions.

Sodicity problems occur when sodium occupies about 15% or more of the soil's exchange sites, but this critical value can be less on some soils. The **exchangeable sodium percentage (ESP)** soil test uses another cation to displace sodium from exchange sites into the soil solution. The solution is then extracted and analyzed for sodium and other cations. The ESP is calculated as follows:

$$ESP = \left(\frac{Na_{exch}}{CEC}\right) \times 100$$

where [Na_{exch}] is the concentration of exchangeable sodium (cmol kg^{-1}), and CEC is the **cation exchange capacity** expressed as centimoles of charge per kilogram of soil (cmol$_c$ kg^{-1}). Because calcium and magnesium are likely to be the only other dominant exchangeable cations on the exchange complex, the denominator of the ESP expression is often replaced with the term; ([Na_{exch}] + [Ca_{exch}] + [Mg_{exch}]).

Sodic conditions adversely affect the soil's physical properties. Sodium-influenced aggregates readily disperse and lower the soil's permeability to air and water. This lowers carbon dioxide diffusion out of the root zone, and its accumulation can retard root growth. Oxygen replenishment is similarly interrupted and all aerobic processes suffer. Water drainage from dispersed soils slows greatly, causing affected areas to remain wet. Since the impermeable layer of sodic soils is often the surface layer, water entry is retarded and irrigation inefficiency, runoff, and erosion increase.

Sodium causes dispersion because it reduces the colloid-to-colloid attraction necessary to aggregate soil particles (Fig. 11-1). Cations with a high charge density (for example Ca^{2+}) bind closely to colloid surfaces, a condition necessary to promote interparticle attraction and flocculation. Sodium has a low charge density and it hydrates, both features that cause it to bind loosely to colloids. The tendency of

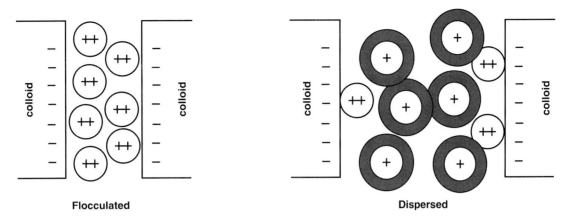

Figure 11-1. Ions that are tightly attracted to charged soil surfaces (Ca^{2+}, for example) allow interparticle forces to flocculate colloids. Weakly attracted ions (Na^+, for example) with their shells of hydrated water reside further away from charged surfaces. This enlarges the distance between colloids and repulsive forces keep the colloids dispersed.

sodium to hydrate attracts water around the colloids, causing them to separate ever so slightly but enough to weaken their attraction. The small residual charge of the loosely bound sodium layer then repels other similarly configured colloids.

In addition to having poor physical conditions, the pH of sodic soils characteristically exceeds 8.5 and often reaches 10 or higher. High pH can reduce the plant availability of several nutrients and create a general chemical environment unhealthy for plants. Sodic soils have a high pH because the hydrolysis of sodium carbonate (Na_2CO_3) increases the bicarbonate (HCO_3^-) and hydroxyl (OH^-) concentration as follows:

$$2\,Na^+ + CO_3^{2+} + H_2O \iff 2\,Na^+ + HCO_3^- + OH^-$$

Irrigation water is a common source of sodium, leading to sodic conditions. Therefore, a sodium content analysis should be an important part of standard water quality tests. Some farm, industrial, and municipal wastes can also contribute significant amounts of sodium when used as soil amendments. All such materials should be tested for sodium before being applied to soil.

Soil Degradation by Salinity and Sodicity

Experimental Procedure

Part I. Determination of saline and sodic criteria

1. **Preparation of Saturated Soil Paste:** Weigh 100 g of three soils (normal, saline, and sodic) into separate, labeled beakers. Slowly mix small quantities of deionized water with each soil until it reaches the saturation point.

- *At the saturation point, the soil will form a paste that glistens with a film of water on the surface, it flows slightly when the container is tipped, and slides cleanly and freely off the mixing spatula.*

2. **pH Determination:** Immerse pH electrodes directly into the saturated paste, swirl gently to achieve good electrode contact, and read pH value.

3. **Determination of Electrical Conductivity:** For each soil:

a. prepare a Buchner funnel by inserting an appropriate filter, moistening it slightly with deionized water, attaching a clean filter flask, and applying suction to achieve proper seating of the filter.

b. transfer the saturated soil paste to the Buchner funnel and spread it evenly over the entire filter's surface.

c. apply suction until 10–12 mL of filtrate have been collected.

d. determine the electrical conductivity of the filtrate. Your instructor will demonstrate the correct procedure for the instrumentation available. Record your results.

Electrical conductivity meters operate on the principle of measuring the flow of an electrical current through the soil extract. As the salt content of the extract increases, the electrical conductivity of the extract also increases.

4. **Determination of Exchangeable Sodium:** An extracting solution (usually ammonium acetate) is used to displace exchangeable sodium into the soil leachate. An aliquot of this extract can be analyzed for sodium using either flame photometry or atomic absorption spectrophotometry. Since these instruments are not typically included in introductory soil laboratory exercises, your instructor may have these soils analyzed elsewhere and provide that information to you.

Part II. Effect of salts and sodium on water percolation in soils

1. Suspend 3 funnels on a rack and place a beaker beneath each funnel stem. Position a medium-speed filter paper in each funnel (e.g., Whatman no. 2 paper).

2. Place 20 g of a normal (non-saline, non-sodic) soil into each of the funnels.

3. Form the soil in the funnels with a slight depression in the center, label the funnels #1 through #3, then:

- *slowly pour 30 mL of deionized water through the soil in funnel #1.*
- *slowly pour 30 mL of 0.5% $CaCl_2$ (calcium chloride) solution through the soil in funnel #2.*
- *slowly pour 30 mL of 0.5% NaCl (sodium chloride) solution through the soil in funnel #3.*

It is essential that the liquid you add goes **through** the soil. Do not add so much at one time that liquid overflows the depression and runs down the side of the funnel. Allow the soil to soak for 5 minutes. Discard any liquid that drains from the soils.

4. Place a graduated cylinder under each funnel. In small increments, add a total of 75 mL of deionized water to each funnel in such a manner that there is always free water in the soil depression but not so much that an excess overflows down the sides of the funnel.

5. For each funnel, monitor and record the time required for each successive 5 mL portion of drainage water to be collected in the graduated cylinder.

6. Plot the rate of water drainage for each soil condition (Note: Your data is best plotted on semi-log graph paper with time units placed on the log axis and drainage volume on the linear axis). Answer the accompanying questions.

Part III. *Effect of soil salinity on seed germination*

1. **Preparation of Saline Soils**

    a. Sieve sufficient normal (non-saline, non-sodic), medium-textured soil through a 4-mm sieve to divide into five 1-kg samples.

    b. Create a salinity range in these samples of 0, 0.1, 0.2, 0.3, and 0.4 percent salt on a dry weight basis by thoroughly mixing 1 kg of soil with either 0, 1, 2, 3, or 4 g NaCl.

    c. Spread samples onto a nonabsorbent surface and use a spray applicator to moisten the soil until the necessary amount of salt has been added. Stir while spraying. Add sufficient water to thoroughly moisten the soil without creating puddling or clumping (approximately 200–300 mL depending on soil type). Add equal amount of water to all five samples. Mix thoroughly.

    d. Store soils in closed, labeled containers at room temperature about two weeks for equilibration.

2. **Germination Test.** This experiment tests five soil treatments, three seed types, and includes two replications, a design that produces 30 germination trials. Responsibility for preparing and/or monitoring these trials may be divided among the class members.

    a. Divide each of the five soils prepared in Part I into six shallow pots. Label pots according to soil treatment. Tap pots on the bench top to settle soil. Arrange pots into six soil treatment sets, each set should have all five of the soil treatments represented.

    b. Into each of the five pots representing one treatment set, place 20 seeds of a high salt tolerant species (barley or bermudagrass) in firm contact with the soil, cover with an additional shallow layer of the same soil, and label according to seed type. Pots should be moistened enough to foster germination and incubated at 25°C. Pots should be covered to reduce water loss and minimize the necessity of adding additional water. Prepare a duplicate trial with a second soil treatment set using the same seed type.

    c. Repeat step 2b two more times, first using seeds with moderate salt tolerance (oats, lettuce, cabbage, or wheat) and then again using seeds with low salt tolerance (whiteclover, red clover, tomato, or celery).

    d. Count and record the number of emerged seedlings daily for two weeks. Check soils for moisture content. If additional water is needed, add the same number of drops of deionized water from a dropper bottle to all treatments.

    e. Prepare a table of experimental observations. Graph the data to show the effect of soil salinity on seedling emergence over time.

3. **Salinity Test.** Determine the salinity of these five soils using the procedure in Part I.

Soil Degradation by Salinity and Sodicity

Name _____ Section _____

Data

Part I. Determination of saline and sodic criteria

	Soil No. 1	Soil No. 2	Soil No. 3
Electrical conductivity, dS m^{-1}			
pH			
Exchangeable sodium, %			
Soil classification (Table 11-1)			
Probable plant response (Table 11-2)			

Part II. Effect of salts and sodium on water percolation in soils. Enter time (mmm:ss) required to accumulate listed volumes of drainage water

Soil	Drainage volume, mL				
	5	10	15	20	25
Normal					
Saline					
Sodic					

Soil	Drainage volume, mL				
	30	35	40	45	50
Normal					
Saline					
Sodic					

Soil	Drainage volume, mL				
	55	60	65	70	75
Normal					
Saline					
Sodic					

Name _____ Section _____

Data

Part III. ***Effect of soil salinity on seed germination (high salt tolerant species)***

	\multicolumn{10}{c}{NaCl added, %}									
Seed type	0	0	0.1	0.1	0.2	0.2	0.3	0.3	0.4	0.4
Day 1										
Day 2										
Day 3										
Day 4										
Day 5										
Day 6										
Day 7										
Day 8										
Day 9										
Day 10										
Day 11										
Day 12										
Day 13										
Day 14										

Data

Part III. Effect of soil salinity on seed germination (moderate salt tolerant species)

Seed type	NaCl added, %									
	0	0	0.1	0.1	0.2	0.2	0.3	0.3	0.4	0.4
Day 1										
Day 2										
Day 3										
Day 4										
Day 5										
Day 6										
Day 7										
Day 8										
Day 9										
Day 10										
Day 11										
Day 12										
Day 13										
Day 14										

Soil Degradation by Salinity and Sodicity

Name _____ Section _____

Data

Part III. **Effect of soil salinity on seed germination (low salt tolerant species)**

_____	NaCl added, %									
Seed type	0	0	0.1	0.1	0.2	0.2	0.3	0.3	0.4	0.4
Day 1										
Day 2										
Day 3										
Day 4										
Day 5										
Day 6										
Day 7										
Day 8										
Day 9										
Day 10										
Day 11										
Day 12										
Day 13										
Day 14										

Part III. **Salinity of soils used in seed germination study**

	Soil Salinity				
Salt, %	0	0.1	0.2	0.3	0.4
EC, dS m^{-1}					

Soil Degradation by Salinity and Sodicity

Name _____ Section _____

Questions

1. How do saline and sodic conditions contribute to soil degradation?

2. What soil tests are required to determine saline and sodic properties in soil?

3. Why is irrigation in dry climates a particular hazard for creating saline and/or sodic conditions in soil?

4. A soil analysis showed a soil to contain 12, 7, and 5 $cmol_c$ kg^{-1} from calcium, magnesium, and sodium, respectively. What is the exchangeable sodium percentage in this soil? Is this soil sodic?

5. Why do sodic conditions cause soils to disperse?

6. Graph the observations from Part II to illustrate differences in percolation between normal, saline, and sodic soils. Summarize these results by explaining the effect that salinity and sodicity have on soil permeability.

Soil Degradation by Salinity and Sodicity

Name _____ Section _____

Questions

7. What significant environmental implications are associated with how soil salinity and sodicity affect permeability?

8. How does salinity affect seed germination?

9. Graph the observations from Part III to illustrate the effect of salinity on seed germination. (Report germination as a percentage of those recorded for the untreated soil.) Summarize these results by explaining the effect of salinity on seed germination.

10. How can the adverse effects on plants of soils with high soluble salt concentrations be overcome?

11. Describe the connection between global water demand, water quality, and soil degradation.

Exercise 12: Soil Testing for Available Phosphorus

> *The chemical condition of a soil is not readily determined except by laboratory investigations called soil tests. Soil tests are analyses that determine the content of targeted chemicals in soil. Soil tests provide information useful to assist managers in evaluating soil properties like nutrient supplying capacity, fertilizer requirements, and potential contamination by pollutants. As increasing amounts of chemical application and pollution occur in the environment, soil tests will continue to provide an important management tool.*
>
> *Exercise Goal: In this exercise you will use the soil test for plant-available phosphorus as a model of other tests. Phosphorus exists in many forms in soil with only a small amount in a plant-available form at any one time. Phosphorus is tested for by first using a chemical extractant designed to separate the portion of the total content in the soil that will become available to plants. Extractable levels are correlated to recommendations through research.*

Chemical tests have been developed for practically all soil components. Scientists rely on these analyses to provide information about the form, concentration, and activity level of such things as nutrients, pollutants, enzymes, fertilizers and other amendments, pesticides, and salts. The knowledge gained from these tests is indispensable for good soil management.

The most common of these analyses measures the nutrient-supplying ability of soils as a guide to fertilizer recommendations. This procedure, **soil testing,** consists of four parts: sampling, analysis, interpretation, and recommendation. Soil sampling represents a vital step because the sample collected must be representative of the site. These samples are then submitted to specialized laboratories where they undergo appropriate extraction and analyses. Laboratory chemical test values are then compared to research data and used to rank the soil according to its ability to furnish nutrients (Fig. 12-1). The soil test level describes the portion of the crop's nutrient requirement supplied by the soil. Finally, the difference between a soil's nutrient-supplying capacity and a crop's need is provided as a fertilizer recommendation.

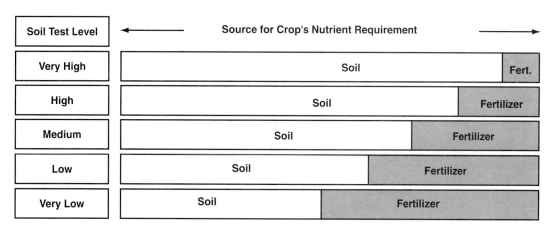

Figure 12-1. As a soil's nutrient-supplying capacity increases, the need for fertilization to meet a crop's nutrient requirement is decreased.

Soil Testing for Available Phosphorus

Routine chemical analysis performed by soil testing laboratories include organic matter content, available phosphorus (P), exchangeable potassium (K), magnesium (Mg), calcium (Ca), soil pH, buffer pH, cation exchange capacity (CEC), and percent base saturation (% BS).

Plant availability of P in soils is typically limited by **fixation,** a process whereby available P is complexed into forms with low solubility. Iron and aluminum form low solubility compounds with P in acid soils, while calcium and magnesium do so in basic soils. The fixing capacity of soils differs according to their individual mineralogy. Soils with high P-fixing capacity reduce fertilizer efficiency and will require a high application rate and/or management. Maintaining soil pH at slightly acid values and banding P fertilizer to reduce its contact with soil have proven effective in reducing fixation.

Plants can't utilize every form of soil P, so testing procedures don't evaluate total soil P but try to identify only the portion of P that will become available for plant uptake. To accurately describe the soil's potential P-supplying capacity, determination methods must be related to the ability of plant roots to extract P. This approach relies on the use of a suitable and effective **extracting agent,** a solution that will remove an amount of P from the soil that is correlated to the amount that can become available to plants. If the correct extracting solution is not used, available P can be over- or underestimated. Soil pH levels and mineralogy are especially used to select the appropriate P extracting agent.

Soil P tests are categorized by the extracting agent used to predict plant available P. Tests have used a variety of extracting agents such as deionized water, weak acidic or basic solutions, and P-complexing sinks. The **Bray P_1** extractant, a mix of $0.025\ M$ HCl and $0.03\ M$ NH_4F initially developed by Dr. R. H. Bray of the University of Illinois, is the most widely used P test in acid to neutral soil (predominantly those in the eastern United States). Another P test used on acid soils, the **Mehlich 3** test developed by Dr. Adolph Mehlich from the North Carolina Department of Agriculture, uses an extractant composed of $0.2\ N$ CH_3COOH, $0.25\ N$ NH_4NO_3, $0.015\ N$ NH_4F, $0.013\ N$ HNO_3, and $0.001\ M$ EDTA. The Mehlich 3 extractant is commonly used in the southeastern United States and has the added advantages of extracting other nutrients, including K, Ca, Mg, Mn, and Zn, and of being compatible with automated analysis equipment. Alkaline soils require a basic extractant. The **Olsen** test, developed by Professor S. R. Olsen of the University of Nebraska, uses $0.5\ M$ $NaHCO_3$ and is recommended when soil carbonate levels exceed 2%. Separate fertilizer response calibrations must be used for each extracting method.

The amount of extractable P determined by any soil test cannot, however, be converted to fertilizer recommendations without data from field experiments that determine plant growth response to several rates of fertilizer applied to similar soils with different P test values. This correlation of laboratory test data and field response is called the **soil test interpretation** phase and produces information like that in Table 12-1. The soil test level designation indicates the relative availability of P for crops grown in the soil. A crop grown on a soil testing "very low" in P would be expected to show a large yield response to an application of P fertilizer.

Table 12-1. Typical Bray P_1 soil test levels correlated to probable yield response from fertilizer additions.

Bray P_1 Soil Test Level	Extractable Soil P		Ability of soil to supply P	Probable response of corn and soybeans to P fertilizer
	mg kg^{-1}	lb acre^{-1}		
Very high	40+	80+	Very high	Very low
High	30–40	60–80	High	Low
Adequate	15–30	30–60	Medium	Medium
Deficient	8–15	16–30	Low	High
Deficient	0–8	0–16	Very low	Very high

Phosphorus in the Environment

Environmental protection requires careful management of plant nutrients. Land stewardship, preservation of water quality, and efficient, sustainable production of agronomic, forest, range, and horticultural commodities all benefit when management decisions are based on proper soil tests.

Nutrient enrichment (eutrophication) of surface waters can occur from sewage, stormwater runoff from livestock areas, or soil erosion. Eutrophication causes increased algal and bacterial activity, which can depress oxygen levels enough to kill aquatic animals. Soaps and detergents are the primary source of phosphorus in sewage effluents and can be controlled through use of materials low in phosphorus or by specialized sewage treatment. Runoff from livestock areas can be controlled through site design and use of green filter zones around water resources. Because phosphorus is sorbed so strongly by soil mineral colloids, little enters surface waters from agricultural sites except when carried with soil during erosion. In this case, erosion control and use of soil tests to assess optimum fertilizer needs are the best tools to minimize any adverse environmental impact of using soils for food production.

Experimental Procedure

This procedure demonstrates use of the chemical soil test procedure for plant-available P and the effect of fixation on soil P test levels. The effect of soil type on P fixation requires soil samples to be prepared two months prior to completing the exercise.

Materials

- *Fine sieve, 30-mesh*
- *Buret*
- *Whatman No. 2 filter paper*
- *Carbon black*
- *Shaker*
- *pH meter*
- *5-mL pipette*
- *25-mL volumetric flask*
- *250-mL flask*
- *Test tubes*
- *Spectrophotometer*
- *Spectrophotometric absorption tube*
- *Acid molybdate stock solution (P-B solution): dissolve 75.25 g of ammonium molybdate, $(NH_4)_6Mo_7O_{24} \cdot 4 H_2O$, in 500 mL of distilled water heated to 60°C. Cool the solution and transfer to a 2000-mL volumetric flask, add 1500 mL HCl (sp. gr. 1.19, 37.5%) and bring to volume with distilled water. Store in a glass stoppered brown bottle to which 100 g of boric acid (H_3BO_3) has been added.* **CAUTION: this reagent is highly corrosive.**
- *Dry reducing agent: Aminonaphthol-sulfonic acid (P-C powder): mix 5 g of 1-amino-2-napthol-4-sulfonic acid with 10 g of sodium sulfite (Na_2SO_3) and 292.5 g of sodium pyrosulfite ($Na_2S_2O_5$). Grind the mixture to a fine powder. Store in a cool place in a sealed brown bottle. This dry mix may be stored for up to one year.*
- *Dilute reducing agent (P-C solution): dissolve 16 g of dry reducing agent in 100 mL of distilled water heated to 60°C. Cool and store in brown bottle. Make new weekly.*
- *Bray's No. 1 extracting solution (0.03 N NH_4F in 0.025 N HCl): dissolve 1.11 g NH_4F and 4.16 mL 6 N HCl in distilled water and bring to 1000 mL volume.* **CAUTION: reagent is highly corrosive.**
- *0.5 N $NaHCO_3$ extracting solution: dissolve 42.0 g $NaHCO_3$ in distilled water and bring to 1000 mL volume.*
- *Chloromolybic acid: dissolve 15.0 g ammonium molybdate, $(NH_4)_6Mo_7O_{24} \cdot 4 H_2O$, in 300 mL distilled water. Add 350 mL of 10 N HCl and bring to 1000 mL volume.* **CAUTION: reagent is highly corrosive.**
- *Stannous chloride reducing solution:*

  Stock Solution: dissolve 10 g $SnCl_2 \cdot 2 H_2O$ in 25 mL conc. HCl. Store in a brown bottle.

  Dilute Solution: transfer 3 mL of stock solution to a 100-mL graduated flask and bring to volume with distilled water. Make solution fresh daily. **CAUTION: reagent is highly corrosive.**

Part I. Soil collection and preparation

The soil P test comprising Parts II, III, and IV can be applied to a wide variety of associated studies. Select the method from those listed here.

A. Demonstration of fixation of fertilizer phosphorus

1. Select three soils similar in texture and organic matter content, one medium acid (near pH 5.5), one slightly acid (near pH 6.5), and one slightly basic (near pH 7.5).

- *Soil selection is critical to illustrate P fixation. Greater fixation occurs in soil with a Bray P_1 level <15 mg kg^{-1} (low or very low soil test). If soil of <15 mg kg^{-1} P or soils of these pHs are not available, several months prior to use add sufficient limestone to a medium acid soil to achieve additional samples with pH levels near 6.5 and 7.5.*

2. To one-half of each sample, thoroughly mix in a P fertilizer material at a rate of 50 mg P kg^{-1} soil (50 ppm P), moisten, and incubate in a warm environment for two months. If a subsoil is used, apply P at 100 mg kg^{-1}). Air dry immediately prior to class use.

3. The remaining half of each soil should likewise be moistened, incubated, and air dried but receive no P fertilization. These preparatory steps create six samples on which the class can collectively perform the following P extraction and analysis.

B. Phosphorus location in soils in reference to environmental quality

1. Collect soil samples from the surface (0–6 inches) and subsurface (6–12 inches) of sites susceptible to various degrees of erosion and/or maintained under various phosphorus fertilizer or manure treatments. Air dry samples for storage.

2. Collect information on site management, erosion, and leaching potential of the soil types sampled and on P contamination problems of nearby surface waters.

Part II. P extraction (select the suitable extraction procedure and use it on duplicate samples)

In this step, extracting solutions remove a portion of the total P in the soil. *Note: Clean glassware is absolutely essential for P analysis; traces of soap residue can lead to large errors.*

A. Extraction of P from acid soils

1. Measure 2.0 g of finely sieved soil into a clean, 250-mL flask.

2. Add 20 mL of Bray's No. 1 extracting solution (0.025 N HCl in 0.03 N NH_4F) from a buret. Stopper the flask and shake continuously for exactly one minute.

3. Filter the solution through a dry, Whatman No. 2 filter paper into a clean, dry, test tube.

B. Extraction of P from neutral, alkaline, or calcareous soils

1. Add 5 g finely sieved soil, 1 teaspoon of carbon black, and 100 mL of 0.5 N $NaHCO_3$ extracting solution to a clean, 250-mL flask.

2. Stopper and shake mixture for 30 minutes on a shaking device.

3. Filter mixture through a Whatman No. 2 filter paper into a clean, dry, test tube. (Add more carbon black if necessary to obtain a clear filtrate).

Part III. Color development

In this step, the P extracted in Part II is complexed into a chemical form that has a blue color. The amount of P extracted is indicated by the intensity of the blue color.

A. Acid soils

1. Transfer a 5-mL aliquot of the filtrate to a test tube.

2. Add 0.25 mL acid molybdate solution (P-B solution) and 4.5 mL of distilled water. Shake to mix.

3. Add 0.25 mL dilute reducing agent (P-C solution). Allow color to develop 10 minutes, then measure absorbance as directed in Part IV.

B. Neutral, alkaline, or calcareous soils

1. Transfer a 5-mL aliquot of the filtrate to a clean, 25-mL volumetric flask.

2. Slowly add 5 mL chloromolybdic acid solution to the flask.

3. After CO_2 effervescence has ceased, shake flask gently and fill flask nearly full by washing the sides with deionized water.

4. Add 1 mL stannous chloride reducing solution and then enough deionized water to bring the flask to volume.

5. Mix contents of the flask, allow color to develop 10 minutes, then measure absorbance as directed in Part IV.

Part IV. Measuring absorbance of colored solution

This step uses a spectrophotometer to convert color intensity to an absorbance reading. The absorbance reading is further converted to a concentration by using a standard curve. The standard curve was developed by determining the absorbance of solutions colored with a known concentration of phosphorus.

1. Rinse a spectrophotometric absorption tube three times with deionized water and once with some of your colored solution. Fill the absorption tube approximately 3/4 full with colored solution.

2. Wipe off any traces of water or fingerprints from the outside of the absorption tube with a soft tissue, place it into the spectrophotometer and note the reading on the absorbance scale.

- *Note: The instrument must be calibrated according to directions before taking your reading. Set wavelength at 660 nanometers (660 nm).*

3. Compare your absorbance value with those on the standard curve and convert to soil concentration using the following formula:

$$mg\ kg^{-1}\ P\ in\ soil = (mg\ L^{-1}\ in\ solution)(dilution\ factor)\left(\frac{L\ of\ extract\ used}{kg\ of\ soil}\right) \qquad Eq.\ 12\text{--}1$$

where: mg L^{-1} in solution = reading taken from standard curve, dilution factor = number of times the aliquot of extracted filtrate was diluted; i.e.,

$$\frac{5 + 0.25 + 4.5 + 0.25}{5} = 2\ (for\ acid\ soils) \quad or \quad \frac{25}{5} = 5\ (for\ alkaline\ soils) \qquad Eq.\ 12\text{--}2$$

and $\frac{L\ of\ extract\ used}{kg\ of\ soil}$ = another dilution factor that relates the P concentration in the sample back to the original P concentration in the soil; i.e.,

$$\frac{20}{2} = 10\ (for\ acid\ soils) \quad or \quad \frac{100}{5} = 20\ (for\ alkaline\ soils) \qquad Eq.\ 12\text{--}3$$

Soil Testing for Available Phosphorus

Name _____ Section _____

Data

	Sample 1	Sample 2
1. Sample Identification	_____	_____
2. Weight of beaker + soil, g	_____	_____
3. Weight of beaker, g	_____	_____
4. Weight of soil, g	_____	_____
5. Spectrophotometer reading, absorbance	_____	_____
6. Solution P concentration, ppm (from standard curve)	_____	_____
7. Dilution factor	_____	_____
8. Volume of extract (mL) ÷ weight of soil (g)	_____	_____
9. Soil test level, mg kg^{-1} (line 6 × line 7 × line 8)	_____	_____
10. Soil test level, kg ha^{-1}	_____	_____
11. Soil test level, lb acre^{-1}	_____	_____
12. Soil test level, average	VL L M H VH	

Soil Testing for Available Phosphorus

Name _____ Section _____

Data

Obtain data from other class members to fill in this table and determine the percent of added fertilizer P that has been fixed by the three soils.

For soils prepared according to Part I-A.	Medium Acid Soil	Slightly Acid Soil	Slightly Basic Soil
1. Soil P test level, mg kg^{-1} (unfertilized soil)	_____	_____	_____
2. Soil P test level, mg kg^{-1} (fertilized soil)	_____	_____	_____
3. Increase in available P level due to fertilizer (line 2 − line 1), mg kg^{-1} P	_____	_____	_____
4. Percent of fertilizer P detected as available, (line 3 ÷ 50 mg kg^{-1}) × 100	_____	_____	_____
5. Percent of fertilizer P fixed (100 − line 4)	_____	_____	_____

Write an evaluation of this study that compares phosphorus fixation among the three soil types.

Soil Testing for Available Phosphorus

Name _____ Section _____

Data

For soils collected according to Part I-B.	Site 1	Site 2	Site 3	Site 4	Site 5	Site 6
1. Soil P test level—surface, mg kg^{-1}						
2. Soil P test level—subsoil, mg kg^{-1}						
3. Erosion potential (high, medium, low)						
4. P fertilizer application rate (high, medium, low)						
5. P level in nearby surface waters (high, medium, low)						
6. Leaching category of soil (high, medium, low)						
7. Potential of soil for furnishing P to environment (high, medium, low)						

Write an evaluation of this study that compares the potential of these soils for furnishing P to the environment.

Name _____ Section _____

Questions

1. Soil tests for phosphorus rely on extracting agents. Describe the role of the extracting agent in estimating the amount of P available to plants.

2. Why don't phosphorus soil tests analyze for total soil phosphorus content? What type of phosphorus is extracted?

3. Why is the plant availability of soil phosphorus sensitive to soil pH? What pH values limit phosphorus availability?

4. Why do laboratory soil test values have to be correlated to fertilizer rate experiments to predict fertilizer need?

5. How can soil tests promote protection of soil and water resources?

Soil Testing for Available Phosphorus

Name _____ Section _____

Questions

6. Convert your soil test level from lb P/A to lb P_2O_5/A (or mg kg^{-1} P to mg kg^{-1} P_2O_5) using the equation: percent P_2O_5 = percent P × 2.3. Show how the factor 2.3 is determined. Where is the term P_2O_5 used?

7. What is eutrophication? Which is a more serious environmental hazard for phosphorus, erosion or leaching? Why?

8. What if the response to added phosphorus fertilizer was lower than the soil test predicted? What soil management practices might be adopted that would increase the response?

9. List the name, address, phone number, and Web address of at least two soil testing facilities available to you.

Exercise 13: Soil Organic Matter and Chemical Sorption

> *The organic fraction of soil, while usually quite small, has an impact on soil biological, physical, chemical, and environmental processes far out of proportion to its quantity. As we learn more about this complex component of the soil ecosystem, two features are becoming quite apparent: (1) organic matter contributes significantly to high soil quality, and (2) organic matter in agricultural soils is declining.*
>
> *Exercise Goal: This exercise illustrates a technique for measuring the amount of organic matter in soil by a combustion procedure. It also includes an analysis of soil cation exchange capacity with the ability to determine the contribution of organic matter to this important soil property. The contribution of organic matter to aggregate stability is demonstrated. Finally, the importance of organic matter as the dominant sorbing surface for pesticides is illustrated.*

Soil organic matter is not a single, readily definable entity but includes a heterogeneous set of entities. Broadly interpreted, it includes all carbon-based components found in soil. These would include a biotic, or living, component comprised of plant roots and a myriad of micro- and mesoflora and fauna. The nonliving segment represents material of plant, animal, or human origin in various stages of decomposition. Carbon compounds in the soil ecosystem face three fates: (1) oxidation and return of the carbon to the atmosphere as carbon dioxide, (2) assimilation into biomass, or (3) incorporation into humus.

Although organic matter is a minor component of soils (soils typically contain 1%–5% organic matter), it performs and influences many essential functions in the ecosystem far out of proportion to the small quantities present (see Fig. 4-1). Practically all physical, chemical, and biological processes in soil show some dependence on organic matter. Because of this importance, organic matter has earned respect as the "lifeblood of soils" and warrants our investigation.

The origin of most soil organic matter is plant tissue. Soil organisms decompose plant tissue and synthesize from it a dark, amorphous colloid called **humus.** Humus is the active component of soil organic matter and enhances water retention, nutrient adsorption, aggregate stability, and pesticide adsorption, to identify a few key contributions. A simplified picture of humus formation from plant residue shows the involvement of enzymatic oxidation by microbes.

$$\text{Plant residue (Reduced carbon)} + O_2 \xrightarrow{\text{Enzymatic oxidation}} CO_2 + H_2O + \text{Energy (Microbial)} + \text{Humus}$$

Humus is a relatively stable form of soil carbon, but it too can eventually be lost as microbes slowly oxidize the energy stored in its structure. A soil having 4% organic matter contains up to 180 million kilocalories of potential energy per acre furrow slice, an equivalent heat value of 25 tons of coal. This energy is released slowly as microbes oxidize soil organic matter to support their growth. Microbial decomposition of humus operates continuously in nature, but slowly enough that scientist believe its half-life to be decades, if not centuries. In this exercise, oxidation will be accelerated through use of a high-temperature laboratory combustion.

Organic Matter Content

An important criteria of soil quality is organic matter content. Color provides an estimate of organic matter content because the humus coating on mineral particles darkens the soil. Dark soils typically have chemical, physical, and biological conditions superior to those of light soils. Using color to judge organic matter content can be misleading, however, unless the comparison is made between soils of similar texture. Because of their low specific surface area, coarse-textured soils require less organic matter to look dark than do fine-textured, high specific surface area soils.

Accounting for soil organic matter content is an important management consideration. Soil testing labs can provide an accurate determination of soil organic matter content. This value has implications in decisions involving water relations, nutrient application, pH buffering, and especially pesticide management.

Cation Exchange and Organic Matter

Cation exchange takes place on the surfaces of inorganic and organic soil colloids, i.e., clays and humus. Cations are bound to colloids by attraction to the negative charges originating within the colloid's structure. The ions held on these exchange sites are major factors in determining soil chemical and physical phenomena like fertility, acidity, salinity, environmental contamination, and aggregate stability.

The cation exchange capacity of a soil is determined by the amount of clay and humus and the type of clay present. The approximate **cation exchange capacity** (CEC) of individual colloids, measured in $cmol_c$ kg^{-1}, are: montmorillonitic clays, 100; illitic clays, 30; kaolinitic clays, 10; and humus, 200. Humus, though usually present in small amounts compared to clays, can have a significant impact on total CEC by virtue of its own high exchange capacity, which is pH dependent. For example, at pH 5 the charge on humus is 140 $cmol_c$ kg^{-1} while at pH 7 it increases to 200 $cmol_c$ kg^{-1}. Soils with low CEC require more careful management to substitute for this deficiency. However, soils with high CEC may also present management problems associated with high clay content unless a significant proportion of the total CEC originates from the organic fraction.

Contribution of Organic Matter to Aggregate Stability

An important role of soil organic matter is through its impact on binding sand, silt, and clay particles into aggregates. Organic matter stabilizes these aggregates by forming bonds between the particles. The polysaccharide component of soil organic matter is frequently credited with initiating aggregate stability while other organic compounds lend long-term integrity to the aggregates.

Aggregation is necessary to create a favorable pore environment in the soil. Without sufficient porosity, or pores of the appropriate size, plant growth suffers and many soil processes are adversely affected. Aeration and drainage improve with the presence of macropores, while water distribution is more highly correlated to the presence of mesopores. Micropores are principally used for water storage. (See Exercise 4, for a discussion of macro-, meso-, and micropores.)

When subjected to agricultural use for tilled crops, organic matter loss from soils through erosion and oxidation is frequently more than the crop's residue replaces. This net loss of soil organic matter is quickly manifested by the symptoms of poor aggregation. Residue conservation, green manuring with grass or legume cover crops, and the addition of composted materials all promote organic matter conservation.

Pesticide Sorption in Soil

The environmental impact of pesticides is affected by their mobility within the soil profile. The primary key affecting movement through soils is whether the pesticide becomes sorbed onto soil particles or remains in solution.

Organic chemicals like pesticides can be adsorbed and/or absorbed by inorganic and organic soil particles. These two processes have a subtle difference; **adsorption** refers to ions held on the surface of colloidal particles, while **absorption** indicates that the ion entered the soil matrix. Since it is sometimes difficult to establish exactly which process occurs, the term **sorption** is used to include both possibilities.

The tendency for a chemical to sorb to soil can be described numerically with partitioning relationships. The **partition coefficient, K_d,** is the ratio of the amount of pesticide sorbed to soil particles to the amount of pesticide remaining in solution.

$$K_d \ (L \ kg^{-1}) = \frac{\text{Pesticide sorbed, mol kg}^{-1}}{\text{Pesticide in solution, mol L}^{-1}}$$

The more strongly a pesticide sorbs to soil, the less likely it is to move away from the site of application and become an environmental problem. A partition coefficient of 1 would indicate an equal distribution of the pesticide between the solid and solution soil phases. A partition coefficient of 1000 would indicate that 1000 times more pesticide sorbs to soil solids than remains in solution. In most cases, it is preferable to have a large K_d value because it means that a large proportion of the pesticide is sorbed by soil.

Since many pesticides sorb most strongly in the organic fraction, the K_d relationship can be focused on the soil organic matter content through an **organic carbon partition coefficient, or K_{oc},** value:

$$K_{oc} = \frac{K_d}{f_{oc}}$$

where f_{oc} = mass of organic carbon/mass of dry soil.

K_{oc} describes sorption per gram of organic carbon in the soil compared to per gram of soil, as K_d reports. K_{oc} values are useful because soil organic matter, which contains about 50% carbon, is by far the dominant sorbing surface for most pesticides. Much of the variation in sorption properties between different soils disappears if K_d values are replaced by K_{oc} values. In either case, the higher the partitioning coefficient, either K_d or K_{oc}, the less likely it will be that a pesticide will be distributed to unintended sites by the soil solution and get into ground or surface water.

Laboratory Activity

Materials

- 3 soils; low (< 1%), medium (1–3%), and high (> 3%) organic matter content
- Bunsen burners or muffle furnace
- Porcelain crucibles
- Spectrophotometer
- Low power (5× to 10×) binocular microscope
- 3 or 4 mm sieve
- Atrazine K_d solution: 500 µg atrazine L^{-1} in water
- Ohmicron Rapid Assay kit for atrazine (Ohmicron, Newtown, PA)
- 50 mL glass test tubes with Teflon caps
- 100-mL volumetric flask
- Pipette(s) to accurately deliver 200 to 500 µL
- Glass fiber filters (0.8 µ)
- 0.2 N $Cu(C_2H_3O_2)_2$: dissolve 18.2 g $Cu(C_2H_3O_2)_2$ in deionized water and bring to 1000 mL
- 1.0 N $NH_4C_2H_3O_2$: dissolve 77 g $NH_4C_2H_3O_2$ in deionized water and bring to 1000 mL
- Concentrated NH_4OH

Part I. Determination of organic matter content by dry combustion

Dry combustion analysis of organic matter in soils is accomplished by measuring the weight loss in a soil sample following high temperature treatment. Heat oxidizes organic matter to CO_2 and H_2O, which escape from the sample.

Note: High temperature can also cause weight loss from dehydration of minerals or decomposition of carbonates. Thus, this method only approximates soil organic matter content and should not be used on soils containing free carbonates.

$$\underset{\text{(organic matter)}}{C_6H_{12}O_6} + 6\,O_2 + \text{heat} \rightarrow 6\,CO_2 + 6\,H_2O$$

1. Place approximately 5 g each of a low, medium, and high organic matter content soil into separate porcelain crucibles and weigh to the nearest 0.01 gram. These soils should have been previously oven dried and stored in a dessicator.

2. Heat the crucibles to a red color over a Bunsen burner. Stir occasionally to aid complete oxidation of the organic matter. Oxidation is complete when the soil becomes light tan, usually in 1–2 hours. Alternatively, this step can be accomplished using a muffle furnace at 550°C for 24 hr.

3. Cool the samples and reweigh. Determine the loss in weight and calculate percent organic matter.

4. Save the soil for use in Part II.

Part II. Determination of cation exchange capacity

Cation exchange capacity of a soil can be measured by removing the cations present and determining their concentration. However, in a natural state the soil may contain ten or more different cations, making this a difficult task. That approach can be simplified by filling the cation exchange sites with one cation while leaching the others from the soil, then testing for only the one cation. In this procedure (see Figure 13-1), copper (Cu^{2+}) cations are added to fill all the soil's exchange sites. Next, ammonium (NH_4^+) cations are added to displace the copper. After washing the copper from the soil, its concentration is determined by developing the characteristic blue cupric color in a basic solution and comparing the color intensity to standards of known concentration.

Six samples will be tested, using soils from Part I—the low, medium, and high organic matter content soils in both their natural condition and after removal of organic matter. The difference in cation exchange capacity before and after oxidation of the organic matter will illustrate its contribution to the total cation exchange capacity of these soils. Label containers and use care to keep the samples identified.

1. Place 1.0 g of each dry soil sample into six separate, small, clean flasks. Add 10 mL 0.2 N $Cu(C_2H_3O_2)_2$ to each flask and mix by swirling for 1 minute. In this step copper cations are displacing all other cations on the exchange sites.

2. Prepare six funnels and filter papers to drain into beakers.

3. Pour the soil and solution into the funnel, washing any remaining soil into the funnel with deionized water in a wash bottle.

Step 1. Copper saturation.

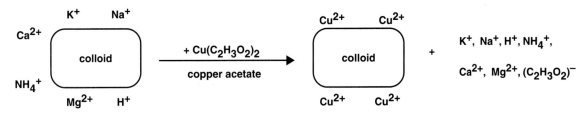

Step 2. Copper extraction and color development.

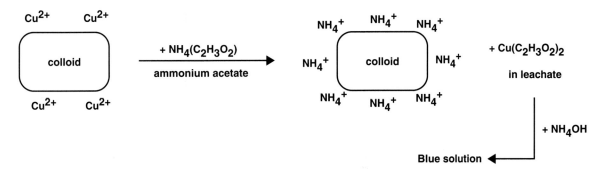

Figure 13-1. Cation exchange capacity determination begins by filling all exchange sites with copper, then displacing the copper with ammonium ions, and finally, testing for the amount of copper removed from the soil.

4. Wash the excess copper acetate from the soil by slowly pouring 30 mL of deionized water through the soil, letting each aliquot soak into the soil before adding more.

5. Discard all filtered solutions and place a clean, graduated cylinder under each funnel.

6. Add three 5-mL portions of 1.0 N $NH_4C_2H_3O_2$ slowly to the soil in the funnel, letting each aliquot soak into the soil before adding more. In this step ammonium cations are displacing copper cations from the exchange sites.

7. Leach deionized water through the soil until a total of 20 mL of solution is in the graduated cylinder. The copper ions that had occupied exchange sites are now washed into the cylinder.

8. Add 5 mL concentrated NH_4OH to the graduated cylinder. As the solution becomes basic, the blue cupric color will develop in proportion to the amount of copper present.

9. Determine the copper concentration by comparing the blue color of the sample to the blue color of standard copper solutions prepared by your instructor. This comparison can be done visually or more precisely with a spectrophotometer at 540 nm and a standard curve.

10. Record the copper concentration and calculate the cation exchange capacity of each soil.

Part III. Contribution of organic matter to aggregate stability

Aggregate stability can be illustrated by observing aggregates rapidly immersed in water. Water enters the aggregate by capillarity from all sides, displacing the air in the pores toward the center of the aggregate. As the air pressure increases, it either (1) forces air bubbles out of the aggregate or (2) exceeds the strength of binding agents in the aggregate, causing it to "explode." Weakly aggregated soils disrupted in this manner will puddle when wet or crust when dry, creating zones where air and water movement are limited.

1. Collect samples from several sites illustrating differences in soil organic matter content. Sieve aggregates and retain those larger than 3 or 4 mm. Air dry.

- *A good comparison is with soil from a cultivated site and an adjacent fence row or similar noncultivated site. Collect soils of similar clay content but varying in color.*

2. Place a shallow dish of water under a low power (5× to 10×) binocular microscope. Water depth must be greater than the aggregate's diameter.

3. Using tweezers, place an aggregate into the water and observe what happens. Stable aggregates will remain intact and air bubbles can be seen escaping. Unstable aggregates will swell and "explode."

4. Repeat with other soils. Compare aggregate stability with organic matter content.

Part IV. Effect of organic matter on atrazine sorption

Source: Basta, N. T., Pesticide Adsorption and Leaching in Soil: An ELISA Laboratory Experiment. J. Nat. Resour. Life Sci. Educ. 24:23–26, 1995.

The fate of the pesticide atrazine applied to soil is of primary environmental concern and is linked to the organic matter content of a soil. Sorption of atrazine by soil components, primarily organic matter, reduces its solution concentration and lessens potential leaching hazards. In this exercise, the tendency for atrazine to be sorbed by a soil will be evaluated.

Atrazine is a widely used selective herbicide for control of broadleaf and grassy weeds in corn, sorghum, rangeland, sugarcane, orchards, and turfgrass sod. It is also used in some areas for selective weed control in conifer reforestation and Christmas tree plantations as well as for nonselective control of vegetation in chemical fallow and noncropland. Depending upon the crop or intended use, atrazine may be applied preplant, preemergence, or postemergence. Atrazine at environmentally relevant pHs will primarily act as a nonpolar organic molecule. At low pH, hydrolysis of atrazine can occur with conversion to a more sorptive compound.

Perform the following procedure on three soils: one each with low, medium, and high organic matter content.

A. Atrazine sorption

1. Place 10 g of air-dried soil into a 50-mL test tube.
2. Add 20 mL of the atrazine K_d solution to each tube, stopper, and shake for 15 min.
3. Filter the sample through a glass fiber filter.
4. Pipette 1 mL of the filtrate into a 100-mL volumetric flask and dilute to volume with deionized water.
5. Determine the amount of atrazine in the flask using the ELISA procedure described in Part IV-B.
6. Record these values and calculate the amount of atrazine sorbed and a K_d value for each soil.
7. Repeat Steps 1–6, except use 20 mL of deionized water instead of the atrazine K_d solution in Step 2. These samples are needed for a negative control that does not contain atrazine.

B. ELISA procedure using the Ohmicron Rapid Assay kit for atrazine

1. Place 200 µL of the sample to be tested in the test tube provided in the kit.
2. Add 250 µL of atrazine enzyme conjugate to the test tube.
3. Pipette 500 µL of mixed Magnetic Particle solution into each test tube and vortex or shake for 15 s.
4. Incubate at room temperature for 15 min.
5. Place test tubes into the magnetic separator for 2 min. With tubes still in the rack, decant the solution from the tubes and drain tubes on absorbent paper.
6. Add 1 mL of washing solution and allow to stand for 2 min. With tubes still in the rack, decant the solution from the tubes and drain tubes on absorbent paper.
7. Repeat the washing in Step 6 one more time.
8. Remove the tube rack from the magnetic separator and add 500 µL of color reagent to each tube. Shake and incubate at room temperature for 20 min.
9. Add 500 µL of stopping solution.
10. Determine the absorbance of the colored solution at 450 nm within 15 min of adding the stopping solution.

11. Calculate percent B/B_o values for samples, where B is the sample absorbance and B_o is the absorbance of a negative control that does not contain atrazine.

12. Determine atrazine concentration from (B/B_o) versus log atrazine calibration plots constructed from atrazine standard solutions provided in the ELISA kit.

13. Disposal of Atrazine: 1) Contact the OSHA-EPA compliance group for assistance in meeting EPA disposal guidelines at your university, or 2) if available, make arrangements with the university's farm superintendent to pickup excess atrazine for use or disposal.

Soil Organic Matter and Chemical Sorption

Name _____ Section _____

Data

Part I. Determination of organic matter content by dry combustion

	Soil Organic Matter Content		
	Low	Medium	High
1. Weight of soil and crucible before combustion, g	_____	_____	_____
2. Weight of crucible, g	_____	_____	_____
3. Weight of soil, g	_____	_____	_____
4. Weight of soil and crucible after combustion, g	_____	_____	_____
5. Weight lost (organic matter), g	_____	_____	_____
6. Organic matter, %	_____	_____	_____

Soil Organic Matter and Chemical Sorption

Name _____ Section _____

Data

Part II. Determination of cation exchange capacity

	Low Organic Matter		Medium Organic Matter		High Organic Matter	
	Natural	Combusted	Natural	Combusted	Natural	Combusted
1. Copper concentration in leachate, $cmol_c$ Cu^{2+}	_____	_____	_____	_____	_____	_____
2. Cation exchange capacity, $cmol_c$ g^{-1}	_____	_____	_____	_____	_____	_____
3. Cation exchange capacity, $cmol_c$ kg^{-1}	_____	_____	_____	_____	_____	_____
4. Exchange capacity attributable to organic matter, $cmol_c$ kg^{-1}	_____		_____		_____	
5. Percent of cation exchange capacity attributable to organic matter	_____		_____		_____	
6. Percent of cation exchange capacity attributable to mineral colloids	_____		_____		_____	

Soil Organic Matter and Chemical Sorption

Name _____ Section _____

Data

Part III. *Contribution of organic matter to aggregate stability*

Soil Description	Clay content* Low, medium, high	Organic matter content 1 = light 5 = dark	Relative aggregate stability 1 = highly unstable, quickly explodes 7 = highly stable, persistent integrity
1.	L M H	1 2 3 4 5	1 2 3 4 5 6 7
2.	L M H	1 2 3 4 5	1 2 3 4 5 6 7
3.	L M H	1 2 3 4 5	1 2 3 4 5 6 7
4.	L M H	1 2 3 4 5	1 2 3 4 5 6 7
5.	L M H	1 2 3 4 5	1 2 3 4 5 6 7
6.	L M H	1 2 3 4 5	1 2 3 4 5 6 7

*Determine using the soil texture by feel procedure in Exercise 2. Designate textures in the lowest tier of the simplified triangle as low clay content, designate textures in the middle tier of the simplified triangle as medium clay content, and designate textures in the upper tier of the simplified triangle as high clay content.

Soil Organic Matter and Chemical Sorption

Name _____ Section _____

Data

Part IV. Effect of organic matter on atrazine sorption

	Soil Organic Matter		
Atrazine Sorption and K_d Determination	Low	Medium	High
1. Absorbance, sample treated with atrazine	_____	_____	_____
2. Absorbance, control sample without atrazine	_____	_____	_____
3. Percent B/B_o (Line 1 ÷ Line 2)	_____	_____	_____
4. Atrazine concentration in diluted filtrate, $\mu g\ L^{-1}$ (from standard curve)	_____	_____	_____
5. Atrazine concentration in undiluted filtrate, $\mu g\ L^{-1}$ (Line 4 × 20)	_____	_____	_____
6. Volume of filtrate, L	_____	_____	_____
7. Atrazine recovered in filtrate, μg (Line 5 × Line 6)	_____	_____	_____
8. Atrazine concentration applied, $\mu g\ L^{-1}$	_____	_____	_____
9. Volume of solution applied, L	_____	_____	_____
10. Atrazine applied, μg (Line 8 × Line 9)	_____	_____	_____
11. Atrazine sorbed by soil, μg (Line 10 − Line 7)	_____	_____	_____
12. K_d (Line 11 ÷ Line 7)	_____	_____	_____
13. Soil mass, g	_____	_____	_____
14. Percent organic matter, from Part I	_____	_____	_____
15. Mass of organic carbon in soil sample (Line 13 × Line 14 × 0.5)	_____	_____	_____
16. f_{oc}	_____	_____	_____
17. K_{oc}	_____	_____	_____

Soil Organic Matter and Chemical Sorption

Name _____ Section _____

Questions

1. Why did the soils change color after extensive heating?

2. What correlation can you draw between combustion losses of organic matter from soils in the lab and organic matter losses caused by soil management practices?

3. With regard to mineral and organic colloids, what differences exist in cation exchange capacity and origin of charge on the colloid?

4. What proportion of the total cation exchange capacity is attributable to the organic matter fraction in each of these soils? If two soils have similar cation exchange capacity, what difference does it make how much comes from the organic matter fraction?

5. Compare the properties of soils with high aggregate stability to those with low aggregate stability. What is the role of organic matter in this comparison?

Soil Organic Matter and Chemical Sorption

Name _____ Section _____

Questions

6. What is a partitioning coefficient?

7. Describe how K_d and K_{oc} differ.

8. What is the correlation between soil organic matter content and the K_d value of a pesticide like atrazine?

9. What is the correlation between the K_d value of a pesticide like atrazine and its movement through soil?

10. Summarize the correlation between soil type, pesticide type, and environmental impact of pesticide use.

Exercise 14: Microbial Decomposition of Organic Materials in Soil

> *The fate of organic material added to soil is to undergo decomposition by a vast microbial community. Decomposition happens in natural ecosystems, fields, gardens, lawns, landfills, and composts. What seemingly represents an inconspicuous, mundane event is, in reality, a vital process in Nature's essential scheme of recycling energy and chemicals. All biological activities, including our existence, depend on the decomposition processes demonstrated here.*
>
> *Exercise Goal: This exercise illustrates carbon's role in energy cycling, nitrogen immobilization and mineralization, and the effects of temperature and C:N ratio on decomposition. You will learn to apply acid-base titration, soil extraction, and colorimetric analysis procedures to the quantification of these principles.*

Microorganisms play significant but inconspicuous roles in many beneficial soil activities, including decomposition. Their role in energy cycling, nutrient transformations, and humus formation are of paramount importance to life on this planet. These actions often go undetected because visual evidence of change occurs too slowly or is obscured within the soil mass. Close examination of decomposing organic matter shows it to be teeming with microorganisms employing a complex series of biochemical reactions to extract life-supporting energy and nutrients. Further along in the decomposition process the original material undergoes changes leading to the formation of soil humus, an essential ingredient of healthy soils. This exercise illustrates decomposition of organic material in soil and has implications directly related to agricultural and environmental management.

Decomposition describes a series of processes that ultimately reduce the complexity of a material. For organic material in soil, this means the constituent parts will either be released or synthesized into new compounds. Nitrogen, phosphorus, and sulfur are examples of nutrients typically released by decomposition, but the same process can also free constituents that might be detrimental to the environment. Released compounds are free to enter into soil chemical reactions, be transported from the site of origin, or be recycled if assimilated by other organisms. Carbon, hydrogen, and oxygen atoms in organic material are initially used for microbial growth or can later be synthesized into humus, but ultimately decompose to two simple molecules: carbon dioxide (CO_2) and water (H_2O). Regardless of their original composition, similarities in breakdown schemes of carbonaceous compounds occur as biological energy is extracted. Eventually, any natural protein, carbohydrate, cellulose, wax, or any organically based product that humans generate degrades to carbon dioxide, water, and small amounts of other residual molecules or ions.

Soil microbes participate in the decomposition of organic compounds to extract energy. Microbes produce specialized proteins, called **enzymes,** that catalyze the transformation of other substances. The energy flow accompanying the transfer of electrons between compounds during enzymatic reactions can be captured by microbes for growth processes. The substance giving up electrons is described as being **oxidized,** while the substance gaining the electron is **reduced.** For example, cellulose molecules in wheat straw represent a reduced carbon compound, meaning it contains stored biological energy. During the acquisition of this energy, microbial processes remove available electrons. Cellulose is first transformed

Microbial Decomposition of Organic Materials in Soil

to a simpler compound, glucose, and then, as electron extraction becomes complete, into carbon dioxide, a metabolic waste product.

$$\underbrace{C_6H_{12}O_6}_{\substack{\text{Glucose}\\ \text{(energy-rich, reduced carbon)}}} + 6\,O_2 \rightarrow \underbrace{6\,CO_2}_{\substack{\text{Carbon dioxide}\\ \text{(energy-poor, oxidized carbon)}}} + 6\,H_2O$$

Since carbon dioxide has no electrons available for any further enzymatic reactions, it is an energy-poor carbon compound. Carbon dioxide gas escapes from the soil to the atmosphere where it can again be reduced (gain electrons) by photosynthesis. By cycling through both decomposition in the soil and photosynthesis, carbon serves as the vehicle of energy flow among heterotrophic and autotrophic organisms.

The decomposition rate of organic materials varies considerably and depends on features inherent to both the residue and the soil environment. Decomposition is regulated by:

Residue Factors:
- *energy availability (readily oxidizable carbon)*
- *nutrient content (especially carbon:nitrogen)*
- *particle size (affects the amount of reactive surfaces exposed)*
- *degree of incorporation (affects soil and residue contact)*

Soil Factors (all affect the rate of microbial activity):
- *temperature*
- *moisture content*
- *oxygen level*
- *chemical composition (nutrient and toxic ion levels)*

The objectives of this exercise are to compare the effect on residue decomposition rates of: (a) energy availability in the various residues, (b) the residue's carbon to nitrogen ratio, and (c) soil temperature. Energy availability in a carbon compound is determined by the types of carbon compounds present. Compounds that require minimal enzymatic action for microbes to derive energy are classed as the most readily metabolized materials. The organic constituents of plants, listed in order of increasingly complex enzymatic requirements during decomposition, are: sugars and amino acids, proteins, hemicelluloses, cellulose, lignin, and a resistant fraction containing fats, oils, and waxes. Composition data in Table 14-1 show that alfalfa contains more readily available energy than does wheat straw, which in turn has more than paper. On that basis, alfalfa is expected to decompose most rapidly.

Table 14-1. Average composition of some organic materials. The distribution of these compounds in the material determines the availability of energy to soil microbes and, hence, their rate of decomposition.

	Protein	Hemicellulose	Cellulose	Lignin
	------------------------------- % -------------------------------			
Alfalfa	18	15	26	10
Wheat straw	2	30	40	12
Paper	—	14	56	24

In addition to furnishing energy, decomposing residue can be a source of nutrients for microbial growth. Nutrients not readily available in the residue must be obtained from soil sources, but at a cost in energy expenditure and decomposition rate. In a nutrient-rich environment, more residue carbon can be diverted to microbial growth, whereas a short nutrient supply means a greater energy expenditure and consequent evolution of carbon dioxide. The distribution of carbon between these two pools is largely determined by the nitrogen pool available to the microbes from the residue. Since the carbon content in most plant residues is about 40%, the variation in nitrogen content can be described by using the carbon to nitrogen ratio (Table 14-2). Residues with low nitrogen content would have wide (high) C:N ratios. Low nitrogen also signifies low protein content, low energy availability, and extended decomposition periods.

Table 14-2. Carbon-nitrogen ratio of some typical soil residues.

Residue	Carbon, %	Nitrogen, %	C:N Ratio
Soil humus	50	5	10:1 (narrow)
Young legumes	40	3.3–2.0	12–20:1
Young grasses	40	2.0–1.0	20–40:1
Manure	40	1.0–0.8	20–50:1
Cornstalks	40	0.7	57:1
Oat or wheat straw	40	0.5	80:1
Tree leaves	40	0.8–0.4	50–100:1
Pine needles	40	0.2	200:1
Woody materials	40	0.1	400:1 (wide)

Soil temperature markedly affects microbial activity through its control of reaction rates. Most microbial activity stops at 0°C but increases rapidly as temperature rise above 10°C. Maximum activity occurs between 25°C and 35°C. Temperatures over 35°C cause a sharp decline in microbial activity.

The addition of organic material initiates several changes in the soil's CO_2 and nitrogen status. The early periods of decomposition are signaled by increases in CO_2 evolution followed by a decline as the available energy becomes limiting (Fig. 14-1). The level of CO_2 evolution provides a direct indicator of microbial activity. By trapping and determining the amount of CO_2 evolved after the addition of organic materials, the relative microbial activity associated with decomposition can be monitored.

During the period of stimulated microbial activity (as noted by high CO_2 formation), the metabolic demand for all nutrients is high. Because sufficient nitrogen is seldom available in the decomposing material, microbes will also utilize nitrogen from soil sources. This process whereby microbes incorporate soil nutrients into their tissue is called **immobilization** and renders the nutrients temporarily unavailable for plants. Later in the decomposition process, as microbial activity declines and microbial tissue decomposes, immobilized nutrients recycle to a mineral form in the soil, a process called **mineralization.** During peak microbial activity, immobilization exceeds mineralization and can create a temporary shortage in the soil, as seen for nitrate in Figure 14-1. These periods of significant immobilization, termed **nitrate depression** in this case, if extended, can stress plant growth and induce nitrogen deficiency. Proper residue management seeks to synchronize the immobilization and mineralization of nitrogen during decomposition with plant needs.

Microbial Decomposition of Organic Materials in Soil

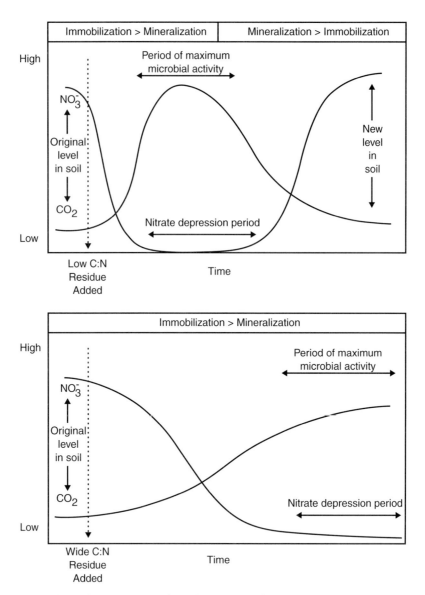

Figure 14-1. When crop residue is added to soil, microbial activity increases as evidenced by an increase in carbon dioxide evolution from the soil and a decrease in nitrate-nitrogen availability in the soil. During the nitrate depression period, all free nitrates are being consumed by microbes and are not available for crop uptake. The rate of decomposition is dependent on the C:N ratio of the residue added. The length of the depression period wil be shorter and level of nitrates in the soil after decomposition will be higher for low C:N residues compared to high C:N residues.

Enhanced microbe populations accompanying residue addition to soil eventually decline as energy becomes more scarce in the decomposing residue. Thus, nitrate depression periods are temporary, existing for periods governed by factors that affect decomposition rate. Mineralization eventually returns all immobilized soil nitrogen plus that which was mineralized from the added residue. Soil nitrogen content following decomposition will be higher than that which existed before residue addition. Residues with the greatest nitrogen content (i.e., those with low C:N ratio) will increase the soil nitrogen level the greatest amount after decomposition.

Composting and Microbial Decomposition

Organic materials produced in abundance can become environmental contaminants unless handled appropriately. One nonburn, nonbury method involves **composting,** or storing materials under conditions that promote their decomposition into a useful soil amendment. Material with a wide C:N ratio often decomposes too slowly for direct application to soil. In that case, allowing partial or full decomposition to occur in a compost pile before adding to soil can convert a problematic waste product into a safe soil amendment. Often, nitrogen is added to a compost pile to eliminate delays in decomposition due to excessive nitrogen immobilization. Composting, however, does not reduce the hazards of mineral contaminants. If these compounds were initially present in the residue, they will also be found in the decomposed product.

Experimental Design and Interpretation of Results

This experiment requires greater inputs of time and effort than others in this text, but it also yields greater information. A successful experiment relies on: (1) understanding the objectives and theory, (2) understanding the experimental design, (3) successfully executing the experimental procedures, and (4) correctly interpreting the results.

The objectives and theory of this experiment were described in the previous section. The experimental design tests three organic materials that vary considerably in energy availability and nitrogen content: alfalfa, wheat straw, and paper. These materials are mixed with soil and incubated in sealed containers for three weeks that allow for carbon dioxide evolution to be monitored. The effects of soil temperature are introduced by incubating a soil and alfalfa mixture at both a cool and a warm temperature. The design allows the amount of carbon dioxide evolved and soil nitrate content of each treatment to be measured on a weekly basis. Careful application of quantitative experimental procedures will ensure the most useful database from which to make experimental interpretations.

Treatment Comparisons

Use the following comparisons to aid in the interpretation of this experiment.

1. Temperature effect: Alfalfa residue (warm) and alfalfa residue (cool)

Changes in soil temperature have marked effects on microbial activity. Studies show most microbial activities are maximized between 25 and 35°C. Activity declines steadily at temperatures higher and lower than this range, until finally stopping altogether at 0°C and 60°C. Keep in mind that soil temperatures vary seasonally, diurnally, and with depth. Under ideal conditions, most chemical reaction rates double for each temperature increase of 10°C. Smaller effects are common under field conditions because other factors typically change along with soil temperature. This comparison shows the effect of soil temperature on the rate of mineralization and immobilization. The same residue may be decomposed quite rapidly in warm soils, but cooler soils in fall or spring may slow decomposition considerably.

2. Decomposition of low C:N ratio residue: Alfalfa (warm) and untreated (warm)

Alfalfa contains more readily decomposable carbon materials than do the other residues tested. Also, the higher nitrogen content of this residue reduces soil nitrate immobilization. The CO_2 evolution peak and nitrate depression should both happen rather quickly.

3. Decomposition of wide C:N ratio residue: Wheat (warm) and untreated (warm)

Adding wheat residue (wide C:N) to soil illustrates the nitrate depression effect. The magnitude of this effect is observed by comparing it with the untreated sample to see how much soil nitrate is immobilized. Adding residue with a C:N ratio greater than 26:1 typically causes periods of net immobilization. The wider the C:N ratio of the residue, the greater will be the extent of the nitrate depression period.

4. Decomposition of very wide C:N ratio material: Paper (warm) and untreated (warm).

The very wide C:N ratio of paper signifies a shortage of readily available energy and a shortage of residue nitrogen. As nitrogen is removed from the soil to enhance microbial growth, an extended nitrate depression period results. Carbon evolution from the added residue will initially be less because of the high cellulose and lignin content. Eventually, whenever decomposition finishes, more CO_2 will be evolved from paper than from the residues with lower C:N ratios. This treatment leaves less carbon remaining in the soil to form humus.

5. Comparison of three organic materials: Alfalfa, wheat, and paper

Comparing residues with different C:N ratios demonstrates the effect of this characteristic on carbon dioxide evolution (microbial activity) and soil nitrate levels (nitrate immobilization).

Experimental Procedure

Materials

- The choice of soil is important for a successful experiment. Select a medium-textured soil for easy filtering. Avoid soils high in fertilizer nitrogen as this masks many of the principals or makes them difficult to demonstrate. Also, avoid soil very low in native nitrogen as this tends to even out effects normally seen from materials with different C:N ratios. A soil testing in the range of 50–75 ppm nitrate works quite well.
- Balance
- Measuring spoon, 1 tablespoon
- 4-ounce, wide-mouth Incubation bottles with airtight lids
- Vermiculite
- Labels
- Finely ground organic materials: with narrow (alfalfa), medium (wheat straw), and wide (uninked paper) C:N ratio
- 1.5 N NaOH (sodium hydroxide): dissolve 60 g NaOH in deionized water and bring final volume to 1000 mL. Titrate against a standard acid to determine exact normality.
- High and low temperature incubation chambers
- 1.0 N $BaCl_2$ (barium chloride): dissolve 104.1 g $BaCl_2$ in deionized water and bring final volume to 1000 mL
- Phenolphthalein indicator: dissolve 0.5 g phenolphthalein in 800 mL ethyl alcohol and bring final volume to 1000 mL with deionized water
- 0.5 N HCl (hydrochloric acid): add 41.3 mL concentrated HCl to deionized water and bring final volume to 1000 mL. Titrate against a standard base to determine exact normality.
- Burets
- 0.4 % $CaCl_2$ (calcium chloride): dissolve 4 g $CaCl_2$ in 1000 mL deionized water
- Suction flasks, Buchner funnels, and no. 2 filter paper
- 4 mL pipette
- Antimony sulfate solution: 0.5 g Sb (antimony) metal +20 mL deionized water + 80 mL conc. H_2SO_4. Heating is necessary to get the metal into solution. **Handle cautiously.** Shake before using.
- 0.0125% Chromotropic acid: dissolve 0.5 g 4,5-dihydroxy-2,7-naphthalenedisulfonic acid, disodium salt (chromotropic acid) in a 4.0 kg bottle of concentrated H_2SO_4. **Handle cautiously.**
- 50 mL and 250 mL flasks
- Spectrophotometer
- Standard 100 ppm NO_3 solution: dissolve 0.7221 g pure, dried KNO_3 in deionized water and bring to a final volume of 1000 mL

Part I. Preparation

1. Place 20 g soil into each of five labeled, 4-ounce, wide-mouth incubation bottles. Add 1 tablespoon of vermiculite to each bottle and mix into the soil by stirring. Vermiculite is an inert mineral and is used here to enhance soil aeration. Label each bottle with the residue type, incubation temperature, and experimenter's name and lab section.

2. To two of the bottles add 0.2 g of a ground plant residue with a narrow C:N ratio and mix well. To a third bottle add 0.2 g of a ground plant residue with a wide C:N ratio and mix well. To a fourth bottle add 0.2 g of finely divided paper. Add no residue to the fifth bottle, it serves as an untreated comparison in this experiment.

3. Add sufficient water to moisten the soil in each bottle. Add the same amount of water to each bottle. Avoid overwatering, as doing so would inhibit the aerobic activities required by the objectives of this experiment.

4. In each bottle, set an open 10-mL vial containing exactly 6.0 mL of 1.5 N sodium hydroxide (NaOH). Record the exact concentration (normality, N) and volume (mL) of this base solution on the data sheet. Secure a lid on each bottle, making it airtight. Handle bottles carefully; if any NaOH spills from the vial, the data from that sample is unusable.

5. Place one bottle with the low C:N residue treatment in a low temperature incubation chamber (~5°C). Place the remaining four bottles in the high temperature incubation chamber (~25°C).

Part II. Incubation

Note to instructors: Separate the class into three groups identified according to the length of time their samples will be incubated.

- *Group 1: One week incubation*
- *Group 2: Two weeks incubation*
- *Group 3: Three weeks incubation*

1. After one week of incubation each group will remove the vial of NaOH from each bottle and label according to the specific treatment. Either proceed directly with CO_2 analysis or seal vials airtight and store in the low temperature chamber for later analysis.

2. Group 1 should proceed directly with nitrate analysis of the soil or seal their incubation bottles and store them at a temperature just above freezing to stop further microbial activity. These soils can be analyzed later for nitrate content.

3. Groups 2 and 3 should place another open vial containing 6.0 mL of 1.5 N NaOH in their bottles, reseal the bottles, and return them to their original incubation chamber.

4. After the samples have spent one additional week in incubation, Groups 2 and 3 should remove the vial of NaOH from each bottle and label. Either proceed directly with CO_2 analysis or seal vials airtight and store in the low temperature chamber for later analysis.

5. Group 2 should proceed directly with nitrate analysis of the soil or seal their incubation bottles and store them at a temperature just above freezing to stop further microbial activity. These soils can be analyzed later for nitrate content.

6. Group 3 should place another vial containing 6.0 mL of 1.5 N NaOH in their bottles, reseal the bottles, and return them to their original incubation chamber.

7. After the samples have spent one more additional week in incubation, Group 3 should remove the vial of NaOH from each bottle, label, and proceed directly with CO_2 analysis.

8. Group 3 should proceed directly with nitrate analysis of the soil.

Part III. Carbon dioxide analysis

The NaOH in the open vial serves as a carbon dioxide (CO_2) collection trap. The carbon dioxide is acidic and neutralizes the base (NaOH), forming a salt (sodium carbonate, Na_2CO_3) and water. Since the NaOH trap was intentionally larger than the expected CO_2 output, all of it was not neutralized. By determining how much NaOH remains and comparing it to the initial amount, the amount of CO_2 trapped can be calculated.

$$CO_2 + \times NaOH \rightarrow Na_2CO_3 + H_2O + \times - 2\,NaOH$$

1. Quantitatively wash the contents of each vial into a separate, clean, 250 mL flask using a wash bottle and deionized water. The amount of water used in this step is not critical and should be sufficient to ensure a complete transfer.

2. Add 10 mL of 1 N barium chloride ($BaCl_2$) to each beaker. Barium chloride and sodium carbonate react to form an insoluble barium carbonate ($BaCO_3$) precipitate. The milky white precipitate provides a visual indicator of the amount of CO_2 produced by the various treatments. The sodium hydroxide not neutralized by CO_2 during incubation remains in solution.

$$\times NaOH + BaCl_2 + Na_2CO_3 \rightarrow BaCO_3 \text{ (insoluble)} + 2\,NaCl + \times NaOH$$

3. Add 5 drops of phenolphthalein indicator to each beaker and titrate to the milky white endpoint with 0.5 N hydrochloric acid (HCl). Phenolphthalein is pink when initially added to the basic solution but the color may fade if it becomes adsorbed onto the precipitate. If this occurs, add more phenolphthalein. This confirms that the color change is due to a pH change and not adsorption of the indicator. Record the exact normality and milliliters of acid used to reach the endpoint.

$$NaOH + HCl \rightarrow NaCl + H_2O$$

4. The amount (milliequivalents) of CO_2 trapped in each vial is equal to the amount (milliequivalents) of NaOH neutralized during the incubation. Titration of the NaOH tells us how much NaOH was **not** neutralized during incubation. Record the amount of acid used in Line 12 of the data sheet and calculate Lines 13 through 19.

Part IV. Nitrate analysis-colorimetric procedure

Exercise caution when using materials in this section that are strongly acidic. Both the antimony sulfate and chromotropic acid are essentially concentrated sulfuric acid solutions, and appropriately safe laboratory techniques must be followed when using these materials.

A. Nitrate extraction

1. Add 80 mL of 0.4% calcium chloride, $CaCl_2$, extracting solution to the soil in the incubation jar, cap tightly, and gently shake for 2–3 minutes.

2. Assemble and label a clean suction flask and Buchner funnel. Seat a Whatman No. 2 filter paper by first moistening the paper and then briefly applying suction.

3. Apply suction and slowly pour contents of incubation jar onto the filter. Continue suction and remove as much liquid from the soil as possible. Save the leachate, as it contains the nitrates extracted from the soil. Nitrates can be removed by this simple process because they are anions (NO_3^-) and therefore not bound to any soil colloids. Extraction could have been accomplished with water; however, the calcium solution helps maintain aggregate stability for rapid filtration.

4. Transfer 4 mL of each leachate by pipette to separate, clean, 50 mL flasks.

B. Color development

5. Add 2.0 mL of antimony sulfate solution to each flask. Swirl to mix contents. Antimony sulfate masks chloride interference in colorimetric analysis. Avoid getting this compound on skin or clothes. Wash it off immediately with water.

6. Add 10 mL 0.0125% chromotropic acid and swirl to mix contents thoroughly. Note the time and complete Step 8 within 10 minutes of completing this step.

Be Careful! The chromotropic acid is a strong acid solution. Handle the flask with care as the solution will become quite hot!

The chromotropic acid-nitrate complex is yellow. The intensity of the yellow color indicates the concentration of nitrates.

7. Set the flask in a cold water bath and cool for 2–3 minutes, swirling occasionally.

C. Measure color intensity

8. Prepare a spectrophotometer cuvette by rinsing it three times with deionized water and once with a small portion of solution from the flask. Then add a sufficient amount of your chromotropic-nitrate solution to the cuvette, wipe off any traces of water or fingerprints from the cuvette with a soft tissue, and place it into the spectrophotometer. Read the value indicated on the absorbance scale. Note: The spectrophotometer must be zeroed and adjusted according to directions before taking your reading. Set the wavelength at 430 nm. Prepare a reference standard using 4 mL of extracting solution, 2 mL antimony sulfate solution, and 10 mL chromotropic acid.

D. Convert color intensity to nitrate concentration

9. Record the absorbance value for your sample on Line 20 and the slope of the standard curve on Line 21. Calculate the nitrate concentration in your soil according to the following formula:

$$NO_3 \text{ in soil, ppm} = A \times B \times C$$

Where:

- A = *the absorbance reading from the spectrophotometer*
- B = *the slope of the standard curve; the slope of the standard curve equals the change in NO_3 concentration divided by the change in absorbance*
- C = *(mL of solution leached from soil) ÷ (g of soil), a dilution factor that relates the nitrate concentration of the tested solution back to the original soil, e.g., 80 ÷ 20 = 4*

10. Discard your solutions as instructed and with care because the solution is quite acidic.

Microbial Decomposition of Organic Materials in Soil

Name _____ Section _____

Data

Microbial Decomposition Worksheet—Individual Results

Record only the data for your treatments on this sheet. Use group results to interpret questions.

		Alfalfa	Alfalfa	Wheat	Paper	None
1. Soil treatment, residue type						
2. C:N ratio		___	___	___	___	None
3. Incubation temperature		Warm	Cool	Warm	Warm	Warm
4. Weight of soil, g		___	___	___	___	___
5. Weight of residue, g		___	___	___	___	None
6. Carbon added, mg (assume 40% C)		___	___	___	___	None
7. Rate of residue addition, lb/A		___	___	___	___	None
kg/ha		___	___	___	___	None
8. Normality of base (NaOH), N	Wk#1	___	___	___	___	___
	Wk#2	___	___	___	___	___
	Wk#3	___	___	___	___	___
9. Volume of base in vials, mL	Wk#1	___	___	___	___	___
	Wk#2	___	___	___	___	___
	Wk#3	___	___	___	___	___
10. Amount of base in vial, meq (N × mL)	Wk#1	___	___	___	___	___
	Wk#2	___	___	___	___	___
	Wk#3	___	___	___	___	___
11. Normality of acid (HCl), N	Wk#1	___	___	___	___	___
	Wk#2	___	___	___	___	___
	Wk#3	___	___	___	___	___
12. Acid added in titration, mL	Wk#1	___	___	___	___	___
	Wk#2	___	___	___	___	___
	Wk#3	___	___	___	___	___
13. Amount of acid in titration, meq (N × mL)	Wk#1	___	___	___	___	___
	Wk#2	___	___	___	___	___
	Wk#3	___	___	___	___	___
14. Base neutralized by acid in titration, meq	Wk#1	___	___	___	___	___
	Wk#2	___	___	___	___	___
	Wk#3	___	___	___	___	___

Microbial Decomposition of Organic Materials in Soil

Name _____ Section _____

Data

Microbial Decomposition Worksheet—Individual Results

		Alfalfa	Alfalfa	Wheat	Paper	None
1. Soil treatment, residue type						
15. Base neutralized by CO_2 evolved during incubation, meq (Line 10 − Line 14)	Wk#1	_____	_____	_____	_____	_____
	Wk#2	_____	_____	_____	_____	_____
	Wk#3	_____	_____	_____	_____	_____
16. CO_2 evolved, meq	Wk#1	_____	_____	_____	_____	_____
	Wk#2	_____	_____	_____	_____	_____
	Wk#3	_____	_____	_____	_____	_____
17. CO_2 evolved, mg (22 × Line 16)	Wk#1	_____	_____	_____	_____	_____
	Wk#2	_____	_____	_____	_____	_____
	Wk#3	_____	_____	_____	_____	_____
18. Carbon evolved, mg (0.27 × Line 17)	Wk#1	_____	_____	_____	_____	_____
	Wk#2	_____	_____	_____	_____	_____
	Wk#3	_____	_____	_____	_____	_____
19. Fraction of residue carbon evolved as CO_2, % (Line 18 ÷ Line 6) × 100	Wk#1	_____	_____	_____	_____	_____
	Wk#2	_____	_____	_____	_____	_____
	Wk#3	_____	_____	_____	_____	_____
20. Absorbance (nitrate analysis)	Wk#1	_____	_____	_____	_____	_____
	Wk#2	_____	_____	_____	_____	_____
	Wk#3	_____	_____	_____	_____	_____
21. Slope of standard curve		_____	_____	_____	_____	_____
22. Dilution factor C		_____	_____	_____	_____	_____
23. Nitrate in soil, ppm	Wk#1	_____	_____	_____	_____	_____
	Wk#2	_____	_____	_____	_____	_____
	Wk#3	_____	_____	_____	_____	_____
24. Nitrate mineralized (+) or immobilized (−), ppm (Line 23: Treated − Untreated)	Wk#1	_____	_____	_____	_____	_____
	Wk#2	_____	_____	_____	_____	_____
	Wk#3	_____	_____	_____	_____	_____

Microbial Decomposition of Organic Materials in Soil

Name _____ Section _____

Data

Microbial Decomposition Worksheet—Group Results

Compute averages using the entire class's data and record on this sheet. Use group results to interpret questions.

		Alfalfa	Alfalfa	Wheat	Paper	None
1. Soil treatment, residue type		Alfalfa	Alfalfa	Wheat	Paper	None
2. C:N ratio		_____	_____	_____	_____	None
3. Incubation temperature		Warm	Cool	Warm	Warm	Warm
4. Weight of soil, g		_____	_____	_____	_____	_____
5. Weight of residue, g		_____	_____	_____	_____	None
6. Carbon added, mg (assume 40% C)		_____	_____	_____	_____	None
7. Rate of residue addition, lb/A		_____	_____	_____	_____	None
kg/ha		_____	_____	_____	_____	None
8. Normality of base (NaOH), N	Wk#1	_____	_____	_____	_____	_____
	Wk#2	_____	_____	_____	_____	_____
	Wk#3	_____	_____	_____	_____	_____
9. Volume of base in vials, mL	Wk#1	_____	_____	_____	_____	_____
	Wk#2	_____	_____	_____	_____	_____
	Wk#3	_____	_____	_____	_____	_____
10. Amount of base in vial, meq (N × mL)	Wk#1	_____	_____	_____	_____	_____
	Wk#2	_____	_____	_____	_____	_____
	Wk#3	_____	_____	_____	_____	_____
11. Normality of acid (HCl), N	Wk#1	_____	_____	_____	_____	_____
	Wk#2	_____	_____	_____	_____	_____
	Wk#3	_____	_____	_____	_____	_____
12. Acid added in titration, mL	Wk#1	_____	_____	_____	_____	_____
	Wk#2	_____	_____	_____	_____	_____
	Wk#3	_____	_____	_____	_____	_____
13. Amount of acid in titration, meq (N × mL)	Wk#1	_____	_____	_____	_____	_____
	Wk#2	_____	_____	_____	_____	_____
	Wk#3	_____	_____	_____	_____	_____
14. Base neutralized by acid in titration, meq	Wk#1	_____	_____	_____	_____	_____
	Wk#2	_____	_____	_____	_____	_____
	Wk#3	_____	_____	_____	_____	_____

Microbial Decomposition of Organic Materials in Soil

Name _____ Section _____

Data

Microbial Decomposition Worksheet—Group Results

		Alfalfa	Alfalfa	Wheat	Paper	None
1. Soil treatment, residue type						
15. Base neutralized by CO_2 evolved during incubation, meq (Line 10 − Line 14)	Wk#1	_____	_____	_____	_____	_____
	Wk#2	_____	_____	_____	_____	_____
	Wk#3	_____	_____	_____	_____	_____
16. CO_2 evolved, meq	Wk#1	_____	_____	_____	_____	_____
	Wk#2	_____	_____	_____	_____	_____
	Wk#3	_____	_____	_____	_____	_____
17. CO_2 evolved, mg (22 × Line 16)	Wk#1	_____	_____	_____	_____	_____
	Wk#2	_____	_____	_____	_____	_____
	Wk#3	_____	_____	_____	_____	_____
18. Carbon evolved, mg (0.27 × Line 17)	Wk#1	_____	_____	_____	_____	_____
	Wk#2	_____	_____	_____	_____	_____
	Wk#3	_____	_____	_____	_____	_____
19. Fraction of residue carbon evolved as CO_2, % (Line 18 ÷ Line 6) × 100	Wk#1	_____	_____	_____	_____	_____
	Wk#2	_____	_____	_____	_____	_____
	Wk#3	_____	_____	_____	_____	_____
20. Absorbance (nitrate analysis)	Wk#1	_____	_____	_____	_____	_____
	Wk#2	_____	_____	_____	_____	_____
	Wk#3	_____	_____	_____	_____	_____
21. Slope of standard curve		_____	_____	_____	_____	_____
22. Dilution factor C		_____	_____	_____	_____	_____
23. Nitrate in soil, ppm	Wk#1	_____	_____	_____	_____	_____
	Wk#2	_____	_____	_____	_____	_____
	Wk#3	_____	_____	_____	_____	_____
24. Nitrate mineralized (+) or immobilized (−), ppm (Line 23: Treated − Untreated)	Wk#1	_____	_____	_____	_____	_____
	Wk#2	_____	_____	_____	_____	_____
	Wk#3	_____	_____	_____	_____	_____

Microbial Decomposition of Organic Materials in Soil

Name _____ Section _____

Questions

1. Use two graphs like those in Figure 14-1 and mark the time axes to indicate where week 1, week 2, and week 3 occur based on the group experimental results for the alfalfa (warm) and wheat (warm) treatments.

2. Prepare a graph similar to that in Figure 14-1 showing the CO_2 evolution and nitrate level for three weeks for each of the five treatment comparisons listed on pages 185 and 186.

3. Prepare a caption for each graph produced in Question 2 that interprets the results in terms of decomposition rate, immobilization, and mineralization.

4. Show how the value 0.27 was derived for use in Line 18 of the data sheet.

5. If a soil treated with some residue has a lower nitrate level than the untreated sample after three weeks of incubation, how would you determine if the treated sample is entering or exiting the nitrate depression period?

6. Draw a flow diagram showing how nitrogen moves between three pools—residue, soil, and microbe—over the course of this experiment.

Microbial Decomposition of Organic Materials in Soil

Name _____ Section _____

Questions

7. After a residue has decomposed, the soil nitrate nitrogen level is higher by some degree than prior to decomposition. What factor of the residue is best correlated with this new level of nitrate?

8. What strategies might be employed in some common cropping systems to avoid a nitrate depression stress period?

Exercise 15: Soil Survey Reports

The soil survey report is an inventory and evaluation of soils in the survey area. It contains information valuable to a wide range of people interested in the characteristics and behavior of soil. Soil maps and descriptions allow identification of soil types. Information is available for predicting the response of soil to natural or managed conditions.

Exercise Goal: In this exercise you will learn how to use soil survey reports to gather information about soils in the survey area. You will apply the rectangular survey system as a tool for locating land sites. Knowledge on how to use written soil descriptions and soil maps in the survey are required in this exercise.

Soil survey reports provide scientific information useful to all who interact with soil resources. The soil survey report is a reference manual and atlas about soils located in a particular area, usually a county. The report provides an inventory of soil resources detailed through soil maps, soil descriptions, and tables of properties related to a variety of land uses. Reports will assist farm managers to protect soils from erosion or to promote optimum yields; assist engineers in selecting suitable sites for roads, buildings, ponds, or other structures; and assist environmental planning associated with soil hydrology (surface water flow and internal drainage), septic tank fields, sewage lagoons, and sanitary landfills. Reports describe the soil's suitability for use as recreation sites and aid foresters in managing woodlands; they rate soils according to range site productivity, irrigation applications, cropping suitability, and windbreak establishment. Additional chemical and physical analyses round out the survey's usefulness. Soil survey reports can add to anyone's knowledge of soils.

Preparing Soil Survey Reports

Using knowledge of principal soil characteristics in an area, soil scientists representing the *National Cooperative Soil Survey* identify soil boundaries on aerial maps with the aid of extensive field sampling, site interpretation, and laboratory analyses. This agency is part of the Natural Resources Conservation Service (NRCS) within the United States Department of Agriculture. In most states soil scientists from the state universities assist data collection and interpretation. It is not uncommon to require five to eight years to produce a completed survey report.

Nearly one-half of the counties in the United States have modern soil survey reports available. They are available from county agricultural offices, state land grant universities, NRCS offices, or U. S. Congress representatives.

Land Description Systems

To locate tracts of land on soil maps in a soil survey report, it is necessary to know how to use legal land descriptions. Two systems have been used in the United States, the metes and bounds system and the rectangular survey system.

Land in the original 13 states and in Maine, Vermont, West Virginia, Kentucky, Tennessee, Texas, and Hawaii have been described using metes and bounds. This system used measured distances and directions

Soil Survey Reports

from natural or human-made landmarks to describe the boundaries of land tracts. This system has limited application in modern society and has been replaced by the **rectangular survey system.**

Rectangular parcels of land can be surveyed with the aid of a north-south reference, called a **principal meridian line,** and an east-west reference, called a **base line.** Range lines have been surveyed parallel to the principal meridian every six miles to the east and to the west of this reference. The resulting north- and south-oriented strips of land divided by range lines are called **ranges.** For example, "range one west (R1W)" is the six-mile-wide strip of land immediately west of the principal meridian extending north and south (Fig. 15-1).

Town lines run east and west and have been surveyed six miles apart and parallel to the base line. The strips of land between town lines are called **towns** (Fig. 15-2). The intersection of ranges and towns create a grid of **townships** across the landscape, each area being six miles square. A township is designated by indicating the range and town in which it lies. The shaded portion of Figure 15-1 is a township designated as T3S, R2W.

Townships can be further subdivided into 36 **sections;** each section has a dimension of one mile to a side (Fig. 15-2). Each section, being a square mile, contains 640 acres. Note the serpentine numbering scheme of the 36 sections, as it will prove useful when locating sites on maps in the soil survey.

When site descriptions for less than 640 acres are required, sections can be subdivided according to the scheme in Figure 15-3. Recall that north is always at the top of a map and note how the various parcels of land can be described through the use of one-half and one-quarter designations. The southeast quarter

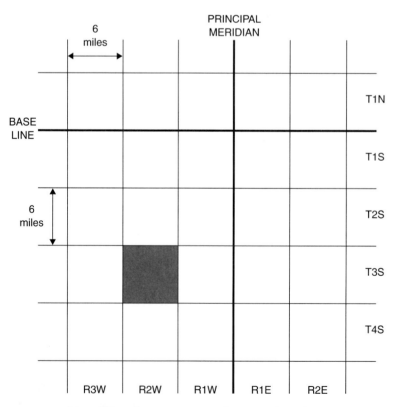

Figure 15-1. The arrangement of meridian, base, town, and range lines in the rectangular survey system. The shaded area represents a township and would be described by: township 3 south, range 2 west (T3S, R2W).

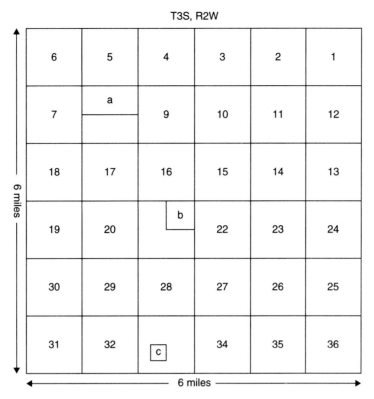

Figure 15-2. A township consists of 36 sections, each a mile square and containing 640 acres.

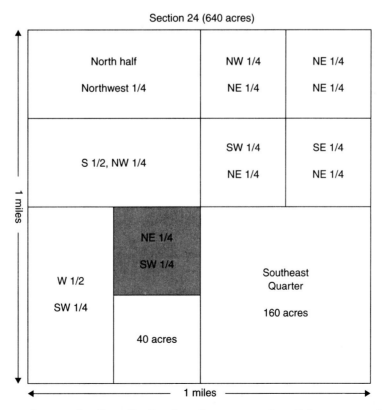

Figure 15-3. Parts of a section can be described using these examples of the rectangular survey system.

(SE 1/4) of this section contains 160 acres. A complete legal description of the 40-acre shaded site in Figure 15-3 would read as: the northeast quarter of the southwest quarter of section 24, township 3 south, range 2 west. Instead of spelling out this description, it can be abbreviated as follows: NE 1/4, SW 1/4, Sec. 24, T3S, R2W. Note that rectangular survey descriptions start with the smallest division and work up to the larger divisions as you read from left to right. Application of this system allows for the description of any size parcel. Irregularly shaped parcels may need several descriptions to include all the land.

Rectangular survey descriptions are included on deeds and other legal descriptions of land. In urban areas, city governments often substitute subdivision plot numbers for describing building sites. Practice using the rectangular survey system by completing Part 1 at the end of this exercise.

Soil Maps

Soils in an area occur in patterns related to geology, landscape features, climate, and native vegetation. Soil scientists will interpret these associations to predict the location of soil types on a landscape. Because individual soils gradually merge with others, accurate soil boundaries can be identified only with the aid of information gathered by site observations and laboratory evaluation. Soil boundaries are then delineated on aerial photographs and each soil is identified with a mapping unit.

Soil **map units** delineate an area dominated by one or several kinds of soil and named according to the taxonomic classification of the dominant soil. Because soils are natural objects, however, a mapping unit will typically include a range of properties characteristic of all natural variability. Mapped areas seldom include only one taxonomic unit. The objective of soil mapping is to separate areas of the landscape into segments that have similar use and management requirements.

Map unit descriptions are included at two levels of detail. The most general soil map units, called **soil associations,** consist of major and minor soils in a distinctive pattern and are named for the major soils. The small scale of general soil maps illustrates soil patterns over large areas but does not allow for farm or field scale identification.

Detailed soil maps are drawn to a larger scale that is suitable for identification of soils at specific sites. Each map unit delineated on a detailed soil map identifies a soil series or sometimes a soil complex. A **soil series** includes those soils with major horizons that are similar in composition, thickness, and arrangement. A soil series can be divided in **soil phases** on the basis of variations in texture, slope, stoniness, salinity, wetness, degree of erosion, or other characteristics that affect soil use (Fig. 15-4). When several soils exist in such an intricate pattern that they cannot be shown separately on a detailed soil map, the area is designated a **soil complex.**

Soil Descriptions

The soil survey contains informative and detailed written descriptions of soils identified in the area. A description exists for each map unit used on detailed soil maps. These descriptions include general facts about the soil, some detailed profile characteristics, and hazards or limitations to use and management of this soil.

Once soils on a site are identified by using the soil maps, soil survey users can refer to the soil descriptions to learn useful features about these soils, including horizon characteristics related to engineering, physical, chemical, and water properties. In addition, the nature and behavior characteristics of the soils are described as they relate to the use and management for crops, pasture, rangeland, woodlands, building sites, sanitary facilities, parks and recreation facilities, wildlife habitat, and

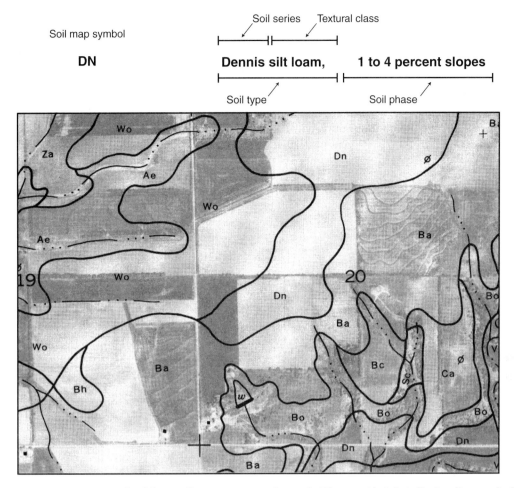

Figure 15-4. Nomenclature used with a soil survey map legend. The partial detailed soil map is from the Wilson County, Kansas, soil survey.

environmental implications. Each soil is assigned to various interpretive groups, either in these descriptions or within a series of tables.

Soil Property Tables

Soil survey reports tabulate many features and properties of soils identified in the area. Information about climate components, distribution of soils, land capability classes, productivity, animal and plant habitat, and recreation or engineering development is typically included. Also, soil surveys include chemical, physical, and engineering indices important to soil scientists, conservationists, environmentalists, engineers, contractors, health officials, and others involved in land use planning.

Digital Soil Surveys

Digital versions of state soil survey manuals and official soil series descriptions are but a portion of the large knowledge base about soils that can be found at http://www.statlab.iastate.edu/soils/nssc/. Visit this site and investigate its abundance of information about soils.

Soil Survey Reports

Name _____ Section _____

Questions

Part I. Rectangular survey descriptions

1. Write the legal description and size for the three labeled areas in Figure 15-2.

   a. _____   _____ acres

   b. _____   _____ acres

   c. _____   _____ acres

Part II. Using the soil survey report

Use a recently published county soil survey report designated by your instructor to answer the following questions. County soil survey reports may be obtained from Agricultural Experiment Stations at the land grant university in your state.

2. Name of county: _____ Date of issue: _____

3. What organization(s) prepared this report? _____

_____

4. Climate facts: Mean annual precipitation: _____

   Average frost-free period (dates or days): _____

   Warmest average daily maximum temperature: Month: _____   Temp: _____

   Coolest average daily minimum temperature: Month: _____   Temp: _____

5. What parent materials are common in this area? _____

_____

Soil Survey Reports

Name _____ Section _____

Questions

6. What native vegetation influenced the development of soils in this area? _____

_____

7. What soil taxonomic orders are found in this area? _____

_____

8. List an example of a soil in each land capability class, its subclass, and major limitation.

	Soil	*Subclass*	*Limitation*
Class I:	_____	_____	None
Class II:	_____	_____	_____
Class IV:	_____	_____	_____

9. Fill in the table for the three most prevalent soil types in the county that are suitable for cultivation. (Note: to be suitable for cultivation, the land capability class must be I, II, III, or IV.) *Your instructor will supply this crop type.

	Soil type	Acres in county	Percent of county	Productivity*, yield per acre	Restrictive feature affecting use for septic tank absorption field.
1.					
2.					
3.					

203

Soil Survey Reports

Name _____ Section _____

Questions

10. Which soil in the county is considered the most productive? _____

    What is the topographical position of this soil? _____

    What characteristics in the description of this soil help to explain its high productivity?

11. Describe these characteristics for the most prevalent soil in the county.

    Depth of A horizon _____   Surface texture _____

    Surface color _____   Depth of B horizon _____

    B horizon texture _____   B horizon color _____

    Special features _____

12. Compare the scale on the general soil map and individual soil maps.

    General map   scale: _____   1 inch = _____ miles

    Detailed map   scale: _____   1 inch = _____ miles

13. What symbols are used on the detailed maps to represent:

    Marsh   House   Rock Outcrop

Soil Survey Reports

Name _____ Section _____

Questions

14. Which soil is most extensive in the following areas? (Your instructor will give you the legal description for this question.)

Legal description	Map unit	Most extensive soil type
a. _____	_____	_____
b. _____	_____	_____
c. _____	_____	_____

15. Assemble a list of features about the soils in Question 14 that pertain to the following uses. Find these features in the various descriptions and tables and describe how they impact the intended use.

   a. The soil in 14a will be used for crop production.

   b. The soil in 14b will be developed for a recreational site.

   c. The soil in 14c will be used for a sanitation facility.

Appendix: Applications of Chemistry to Soil Science

To the Instructor: Many of the exercises in this manual require the application of fundamental chemical concepts and terminology similar to those associated with introductory college chemistry courses. The role of chemistry in soil investigations is so intertwined with their comprehension that this section has been added to ensure maximum understanding. This section can be used as a lesson by itself, or portions can be assigned as supplements to associated exercises.

Part I. Definitions

Atomic weight	A unitless number used to rank atoms according to their composition of neutrons and protons. The atomic weight of an element is obtained from a Periodic Table.
Avogadro's number	$N = 6.023 \times 10^{23}$
Disperse	A chemical or mechanical action to separate particles from each other.
Endpoint	The condition in an acid-base titration when chemically equivalent amounts of acid and base are present; usually designated by a color change of an indicator dye.
Equivalent	Avogadro's number of chemical charges.
Equivalent weight	The mass (gram) of a substance necessary to furnish Avogadro's number of charges.
Filtrate	The liquid that has passed through a filter.
Flocculate	A chemical action that aggregates or combines individual particles into masses of particles.
Leachate	The liquid that percolates through a soil sample.
Mesh	Terminology used to designate sieve size in terms of openings per inch.
Mole	Avogadro's number of molecules; i.e., 6.023×10^{23} molecules of any substance is a mole.
Molecular weight	The mass (gram) of the 6.023×10^{23} molecules composing a mole of a substance. This value can be obtained by summing the atomic weight of all atoms in a molecule.
Normality	A value signifying the concentration of a chemical solution expressed in units of milliequivalents per milliliter (which is the same as equivalents per liter).
Quantitative transfer	A laboratory manipulation that ensures total conveyance of material from one location to another, sometimes accomplished with repeated washings.

Appendix

SI units

An international system of units based on the metric system used in science applications.

Multiples and submultiples of SI units:

10^{12}	tetra	T
10^{9}	giga	G
10^{6}	mega	M
10^{3}	kilo	k
10^{-3}	milli	m
10^{-6}	micro	µ
10^{-9}	nano	n
10^{-12}	pico	p
10^{-15}	femto	f
10^{-18}	atto	a

Solution — Dissolved material in a liquid medium.

Suspension — Small particles dispersed in a surrounding medium.

Valence — A measure of combining capacity as designated by the electrical charge associated with a substance.

Part II. Soil chemistry

This abbreviated periodic table contains elements of particular interest to soil science. Some are essential plant nutrients, others are problem or pollutant elements, some are soil mineral components, and some are useful in laboratory work with soils.

A. For each element on this periodic table, write in the atomic weight, common valence(s), and name (see example for phosphorus).

Soil Science Periodic Table

Legend:
- Atomic weight: 31
- Valence: +5
- Symbol: P
- Name: Phosphorus

IA	IIA	IIIB	IVB	VB	VIB	VIIB	VIIIB			IB	IIB	IIIA	IVA	VA	VIA	VIIA
H																
												B	C	N	O	
Na	Mg											Al	Si	P	S	Cl
K	Ca					Mn	Fe	Co	Ni	Cu	Zn					
Rb	Sr				Mo						Cd					
Cs	Ba										Hg				Pb	

209

Appendix

B. Complete this table for chemical elements important to soil science (see Part A).

Element	Essential nutrient	Plant-available form, if essential	Soil supply source	Reason this element can become an environmental hazard
	Y or N			
	Y or N			
	Y or N			
	Y or N			
	Y or N			
	Y or N			
	Y or N			
	Y or N			
	Y or N			
	Y or N			
	Y or N			
	Y or N			
	Y or N			
	Y or N			
	Y or N			
	Y or N			
	Y or N			
	Y or N			
	Y or N			
	Y or N			
	Y or N			
	Y or N			
	Y or N			
	Y or N			
	Y or N			
	Y or N			
	Y or N			

Appendix

Part III. Volumetric chemistry

A. Exercises using quantitative chemistry concepts require knowledge of the volume and concentration of the liquid in order to express the chemical amount.

$$Volume \times Concentration = Amount$$

or,

$$milliliters, mL \times milliequivalents\ per\ milliliter, meq\ mL^{-1} = milliequivalents, meq$$

or,

$$milliliters, mL \times Normality, N = millequivalents, meq$$

B. At the endpoint of an acid-base neutralization titration, equal amounts of acid and base are present. The conditions at the endpoint can be written using the equations in Part A as follows:

$$milliequivalents_{(acid)} = milliequivalents_{(base)}$$

$$milliliters_{(acid)} \times Normality_{(acid)} = milliliters_{(base)} \times Normality_{(base)}$$

Appendix

C. Two basic computations involving equivalents are needed when working with chemistry problems: either converting equivalents to grams, or *vice versa*. Examples of each conversion using the concept of an equivalent as Avogadro's number ($N = 6 \times 10^{23}$) of charges are given here. For simplified calculations, leave all exponents of ten as the 23rd power.

Example A: Converting equivalents to grams (i.e., chemical amount to mass)

How many grams of sodium are in 2.1 equivalents of sodium?

Step 1. Determine the number of charges in 2.1 equivalents.

(# equivalents) (charges/equivalent)* = # charges

$(2.1)(6 \times 10^{23}) = 12.6 \times 10^{23}$ charges

Step 2. Convert number of charges to number of ions.

(# charges) ÷ (# charges / ion)† = # ions

$(12.6 \times 10^{23}) \div (1) = 12.6 \times 10^{23}$ ions

Step 3. Convert number of ions to mass of ions.

(# ions) (grams / ion) = grams

$(12.6 \times 10^{23})(3.8 \times 10^{-23} \text{ g / Na}^+ \text{ ion})$§ $= 48.3$ g Na$^+$

Example B: Converting grams to equivalents (i.e., mass to chemical amount)

How many equivalents of calcium are in 2 grams of calcium?

Step 1. Convert mass of ions to number of ions.

(gram) ÷ (gram / ion)§ = # ions

$(2) \div (6.6 \times 10^{23}) = 0.3 \times 10^{23}$ ions

Step 2. Convert number of ions to number of charges.

(# ions) (# charges / ion)† = # charges

$(0.3 \times 10^{23} \text{ ions})(2) = 0.6 \times 10^{23}$ charges

Step 3. Convert number of charges to equivalents.

(# charges) ÷ (# charges / equivalent)* = # equivalents

$(0.6 \times 10^{23}) \div (6 \times 10^{23}) = 0.1$ equivalents Ca^{2+}

---

* from definition of Avogadro's number

† from the Periodic Table, i.e., the valence

§ obtained by dividing the mass of one mole of the ion (available on the Periodic Table) by the number of ions in one mole (i.e., Avogadro's number, 6×10^{23})

Appendix

Name _____ Section _____

D. Chemistry Concepts Worksheet

Use the concepts and equations in Parts III-A, B, and C to calculate these answers.

_____ 1. What is the molecular weight of NaOH?

_____ 2. What is the equivalent weight of Na?

_____ 3. What is the equivalent weight of NaOH?

4. How many grams of NaOH does it take to make one liter of:

_____ a. 1 Normal NaOH (written as 1 N NaOH)

_____ b. 0.01 N NaOH

_____ 5. How many equivalents of NaOH are contained in 1 liter of a 1 N solution?

_____ 6. How many milliequivalents of NaOH are in 1 liter of a 1 N solution?

_____ 7. How many milliequivalents of NaOH are in 1 milliliter of a 1 N solution?

_____ 8. How many milliequivalents of NaOH are in 1 milliliter of a 0.01 N solution?

_____ 9. Fifty mL of a 1 N NaOH solution would contain how many milliequivalents of sodium hydroxide?

_____ 10. Fifty mL of a 0.01 N NaOH solution would contain how many milliequivalents of sodium hydroxide?

_____ 11. One milliequivalent of hydrochloric acid (HCl) would dissociate to provide how many milliequivalents of H^+ ?

_____ 12. If 20 mL of 0.01 N NaOH were needed in a titration to reach the endpoint, how many milliequivalents of NaOH would be used?

_____ 13. How many milliequivalents of hydrochloric acid would be neutralized by 20 mL of 0.01 N NaOH?

_____ 14. How many milliequivalents of H^+ from hydrochloric acid would be neutralized by 20 mL of 0.01 N NaOH?

_____ 15. How many equivalents are present in 8 g magnesium?

_____ 16. How many grams of potassium are in 10 milliequivalents of K^+?

Appendix

Part IV. Conversion factors for SI units

To convert Column 1 into Column 2, multiply by:	Column 1 SI unit	Column 2 non-SI unit	To convert Column 2 into Column 1, multiply by:
\multicolumn{4}{c}{Length}			
0.62	kilometer, km	mile, mi	1.61
1.09	meter, m	yard, yd	0.91
0.39	centimeter, cm	inch, in	2.54
\multicolumn{4}{c}{Area}			
0.386	square kilometer, km^2	square mile, mi^2	2.59
247	square kilometer, km^2	acre	4.05×10^{-3}
2.47	hectare, ha	acre	0.405
10.76	square meter, m^2	square foot, ft^2	0.29×10^{-2}
\multicolumn{4}{c}{Volume}			
35.3	cubic meter, m^3	cubic foot, ft^3	2.83×10^{-2}
6.10×10^4	cubic meter, m^3	cubic inch, in^3	1.64×10^{-5}
2.84×10^{-2}	liter, L (10^{-3} m^3)	bushel, bu	35.24
3.53×10^{-2}	liter, L (10^{-3} m^3)	cubic foot, ft^3	28.3
0.265	liter, L (10^{-3} m^3)	gallon	3.78
33.78	liter, L (10^{-3} m^3)	ounce (fluid), oz	2.96×10^{-2}
2.11	liter, L (10^{-3} m^3)	pint (fluid), pt	0.473
\multicolumn{4}{c}{Mass}			
2.20×10^{-3}	gram, g (10^{-3} kg)	pound, lb	454
3.52×10^{-2}	gram, g (10^{-3} kg)	ounce (avdp), oz	28.4
2.205	kilogram, kg	pound, lb	0.454
1.102	megagram, Mg (tonne)	ton (U.S.)	0.907
\multicolumn{4}{c}{Yield and Rate}			
0.893	kilogram per hectare, kg ha^{-1}	pound per acre, lb $acre^{-1}$	1.12
0.107	liter per hectare, L ha^{-1}	gallon per acre	9.35
893	tonnes per hectare, t ha^{-1}	pound per acre, lb $acre^{-1}$	1.12×10^{-3}
2.24	meters per second, m s^{-1}	mile per hour, mph	0.447
\multicolumn{4}{c}{Pressure}			
9.9	megapascal, MPa (10^6 Pa)	atmosphere	0.101
10	megapascal, MPa (10^6 Pa)	bar	0.1
2.09×10^{-2}	pascal, Pa	pound per square foot, lb ft^{-2}	47.9
1.45×10^{-4}	pascal, Pa	pound per square inch, lb in^{-2}	6.90×10^3
\multicolumn{4}{c}{Temperature and Ion Exchange Capacity}			
(9/5 °C) + 32	Celsius, °C	Fahrenheit, °F	5/9 (°F − 32)
1	centimole per kilogram, cmol kg^{-1}	milliequivalents per 100 gram, meq 100 g^{-1}	1